Crab Wars

CRAB WARS

A Tale of Horseshoe Crabs, Ecology, and Human Health

Second Edition

WILLIAM SARGENT

Brandeis University Press
WALTHAM, MASSACHUSETTS

BRANDEIS UNIVERSITY PRESS

Manufactured in the United States of America
Designed by Katherine B. Kimball
Typeset in Galliard

First Brandeis University Press edition 2021
Previously published by University Press of
New England in 2002
ISBN for the Brandeis edition: 978-1-68458-076-7
ISBN for the Brandeis ebook: 978-1-68458-077-4

Library of Congress Cataloging-in-Publication Data
appear at the back of the book

5 4 3 2 1

Contents

Preface to the Second Edition

In the preface to the first edition of this book I asked myself why was I writing about horseshoe crabs when three thousand people had just died in the rubble of the World Trade Center. I discovered that John Steinbeck had asked the same question when he was writing about fish dying in tidepools while Europe was being bombed during World War II. He decided that none of it was very important or all of it was, and I felt the same way.

Today I find myself finishing the second edition of this book in the midst of the worst pandemic in a hundred years, during the most momentous election in my lifetime, and on the day that Denmark decided to euthanize 17 million mink to prevent a new pandemic, for which we might not be able to make vaccines.

Denmark's decision suggests that COVID-19 was not the result of natural selection, but was rather the result of what is called gain-of-function research in ferrets. After all, minks are just larger first cousins of ferrets, with nicer fur coats.

Be that as it may, I could be asking myself the same question. Why am I still writing about horseshoe crabs when over a million people have just died from COVID-19, with almost a quarter of them in the United States alone?

But things have radically changed since 2001. Then I was about the only nontechnical person writing about horseshoe crabs and the public was unaware how important they were becoming to modern medicine.

Today the medical uses of horseshoe crabs are almost common knowledge. Since 2018, the *New York Times*, the *New Yorker*, and the *National Geographic* magazine have all carried major stories about horseshoe crabs in modern medicine, and more recently journalists have written stories about how crucial horseshoe crabs will be in ensuring that both the antibody tests and COVID vaccines will not be contaminated with endotoxins from Gram-negative bacteria—bacteria that are as lethal and ubiquitous as they sound.

But the big change is that in 2003, Dr. Ling Jaek Ding from the National University of Singapore discovered a way to use gene-splicing techniques to replicate the factor in horseshoe crab blood that allows the arachnids to detect and isolate endotoxins. If this method is widely adopted, pharmaceutical companies will not have to rely on a species of crabs that are declining around the world.

But there are debates about the sensitivity and cost of this artificial new product, and to date pharmaceutical companies have been loath to adopt it because only two companies have licenses to produce the recombinant factor and the Food and Drug Administration doesn't want to give them permission to switch to an untried product in the midst of an ongoing pandemic.

So I decided to write this new edition of *Crab Wars* to explore the role that horseshoe crabs will play in this medical saga to rid ourselves of the present pandemic and help us move on to a more healthy world.

Ipswich, Massachusetts
November 4, 2020

Preface to the First Edition

During World War II, John Steinbeck wrote a book about collecting marine animals with his mentor Ed Ricketts. He spoke of taking a tiny colony of corals from a tidal pool and how it wasn't very important to the pool itself. He wrote of Japanese fishing boats dredging up tons of shrimp and how that wasn't very important to the Pacific Ocean. He wrote of the bombs that were then falling throughout the world and how the stars were not moved thereby. He wrote of how all of it is important or none of it was important at all.

I was reminded of those words as I replayed the image of the twin towers of the World Trade Center tumbling into a smoldering pile of dust. What was I doing writing about horseshoe crabs when thousands of people had just died and the airwaves were full of the inciting rhetoric of a holy war?

But as the days unfolded, I gradually realized I was doing something important. I was writing about an animal that has been on this planet three hundred times longer than our own species. I was honoring an animal that has saved a million more human lives than died in the World Trade Center, or will probably die in the aftermath. I was writing about caring for our precious planet that we think we can bomb with impunity to achieve safety from members of our own species—people genetically identical to ourselves. I was trying to show the miracle of the universe, the miracle that all life is related, the miracle that all people on the earth are part of the same contentious family, the reality that we must find a way to live together on our wonderful teeming planet or destroy it for all eternity. In my own way I was praying for that never to happen.

As Steinbeck wrote during the last war that so endangered our planet, "None of it is important or all of it is." I like to think the latter.

Pleasant Bay, Massachusetts
September 14, 2001

Acknowledgments

Many people have helped me write this book. Most have been mentioned in the body of the text. However, I would also like to thank Karen McCullough, Bill Hall, Jim Berkson, and Tina Berger for technical advice; Jim Fair, Frank Germano, and Hoyt and Deborah Ecker for Massachusetts fisheries information; and Scott Hecker of the Massachusetts Audubon Society for general information and kibitzing. Carl Shuster is the grand old man of horseshoe crab researchers. He helped throughout the process of writing this book. I am eagerly awaiting his book.

Norman Wainwright read through the entire manuscript thoroughly. Judge Zobel and Jim Cooper took the time to read over specific chapters. George Buckley was there at the start of our continued explorations of Pleasant Bay. And Troy Hopkins and his students continue the ongoing tradition.

Phil Pochoda, Phyllis Deutsch, Nick King, Scott Allen, and Glenn Ritt have all helped publish previous articles about horseshoe crabs. Betsy Bang, Jack Levin, and John Valois helped with stories of the early days. Jan Nicholls was particularly helpful with recent happenings.

Dr. Chris Miller made suggestions for the latest edition of this book, which was published under the gentle guidance of Sue Ramin and Lillian Dunaj from Brandeis University Press. I can't thank you enough for all your help.

PART I

Early Lessons

Introduction

IN 1956 A HORSESHOE CRAB sludged up and died, not an uncommon occurrence. But this horseshoe crab happened to die in the laboratory of Dr. Frederik Bang, an uncommon scientist who recognized this death was uncommonly like that of rabbits injected with Gram-negative bacteria. That observation led to an exquisite new test for bacterial contamination.

Today, over a million human lives have been saved by the horseshoe crab test, and the processed blood of these animals is worth over $15,000 a quart. It is used to detect infinitesimally small quantities of Gram-negative bacteria, which are as ubiquitous in the natural environment as they are lethal in the human bloodstream. The Food and Drug Administration now requires that every scalpel, drug, syringe, and flu shot be tested with the horseshoe crab derivative called *Limulus* amoebocyte lysate, LAL, or lysate for short.

Producing lysate has become a multimillion-dollar industry headquartered in Boston, Tokyo, Switzerland, and Chicago. There has been intense competition between some of the world's major pharmaceutical conglomerates to develop or acquire subsidiary companies able to produce stable lysate. But little attention has been paid to the unique animals that have made it all happen. Today the industry is plagued by overfishing, dwindling stocks, problems with endangered species, and regulatory uncertainty.

I have tried to tell the story of the horseshoe crab industry through the eyes of the scientists, fishermen, and biotech pioneers involved. In doing so I have traveled the length of the East Coast, interviewed hundreds of people, studied research papers and court documents, and worked inside the industry.

The story is an intensely human one. Like any good story it has rogues and scalawags as well as dedicated scientists—sometimes all wrapped up in the same skin. In many cases I have tried to re-create conversations from long ago. This has not been easy. Few people thought the conversations very memorable at the time, and nobody was taking notes. I have tried to remedy this situation by interviewing everyone involved and comparing their memories of those past events. But people tend to remember details differently. In some cases the participants in the conversations have died, so I have had to rely on the recollections of their wives or colleagues. While I cannot guarantee that these conversations are accurate in every detail, I believe they faithfully reflect the underlying truth and feel of these past events. Where possible I have supplemented people's memories with quotes from courtroom documents, newspaper articles, and meeting transcipts.

Some of the information in this book comes from my own lifelong interest in horseshoe crabs. My involvement with them stems from childhood fascination and has run the gamut from studying the crabs' natural history to writing about their physiology to collecting them for lysate. Each has given me insight into the horseshoe crab industry and the legal, environmental, and economic pressures it faces today.

This is a little-known story that affects people, endangered species, and human health worldwide. It shows some of the unforeseen consequences of biotechnology's powerful new abilities to affect nontarget species and alter the finely tuned balance of natural ecological systems. I believe this story holds important lessons for humanity as we plunge headlong into the rapidly changing world of biotechnology. I hope it shows the importance of regulating these industries, which promise so much but can hurt our planet so severely if we allow them to grow unchecked.

CHAPTER 1

A Day in the Life of a Hunter-Gatherer

IT IS EARLY MORNING. I lie in bed savoring the sounds of songbirds swelling to greet the new day. They reach a crescendo as the first rays of sunlight stream through my window, suffusing the knotty pine walls with a warm golden glow. The calls of crows beckon me outside. They are probably hectoring a great horned owl who has has once again been tardy returning to her nest overlooking the marsh behind our house.

I slip on a bathing suit and tiptoe down the creaky back stairs trying to avoid waking my recalcitrant family. It is a battle I often lose. In my eagerness to rush outside I burst through the screen door, which slams shut behind me.

"Go back to sleep!" bellows my sister from her nearby bedroom.

Outside, the sun is shining on two small rabbits timidly nibbling the dew-drenched grass. They pause to stare with deep dark eyes. A twitch, a flicker, and they bound beneath a thicket of fragrant bayberry already abuzz with honeybees. All that remains are the dark trails of the rabbits' footprints where they ran through the cool wet meadow.

I walk to a row of tomatoes that grow beneath my sister's bedroom. They thrive here, bathed in the first light of day yet protected from the wind. The vines are beaded with shimmering drops of dew caught in the tiny hairs of the succulent leaves. As I brush through the tomatoes, I'm enveloped in the smell of their leaves. It is the tangy odor of the earth, the sun—of summer.

I pluck a tomato and bite into its red, still sun-warmed flesh. Cool juices explode into my mouth and dribble down my bare chest. I am consuming the earth, drinking the sun, breathing in the universe, exhaling the raw materials of life itself.

Lizardlike, I sit on the slope of our white bulkhead absorbing the sun, savoring the tomatoes. Beneath my feet pill bugs roll into perfect spheres, as did their ancestors in the earliest Proterozoic seas. I could not be more in the present, more childlike, more attuned to the rhythms of our ancient, everchanging universe.

"Bill-ly!"

My reverie is instantly broken. My family has arrived en masse to partake of their indoor, to my mind more pedestrian meal. They argue and plan, bicker and gulp great spoonfuls of politically incorrect consumables. I never let on that I have just dined with gods, been drunk with the essence of the universe.

After breakfast, the house starts to fill with the drowsy heat of August. Fat flies blunder into screen doors, and a lone cicada scrapes its strident song from a nearby pine. It is time to get on with the real business of the day, exploration and investigation.

I load an ancient wooden-slatted wheelbarrow with the tools of my trade: crab nets, fishing poles, clam rakes, and wire baskets—best to be prepared for any eventuality.

I pause at the rickety plank bridge that spans the creek running out of the marsh. I have to lie on the bridge to reach the codline that lifts my minnow trap to the surface. Suddenly the trap is alive with thrashing, splashing fish. I pull a few minnows out of the trap and plop them into a bucket of water. They will come in handy if we come across some "snappers," the aptly named young bluefish who have just arrived from their North Atlantic breeding grounds.

Finally I reach the beach and haul in my flat-bottomed plywood skiff. My father built the boat for skimming through the shallow creeks of the marshes behind the islands. Clamped firmly to the transom of the skiff is my pride and joy, a navy blue, three-horsepower Evinrude engine. Brand loyalty is fierce on the bay. You have either an Evinrude or a Johnson outboard, drive a Ford or Chevy station wagon, vote Democrat or Republican. My friends and I have long discussions about the relative merits of each.

I set the choke and pull the cord. A cough, a sputter, a puff of blue smoke, and the battle is finally won. The motor starts and holds a slow, steady rhythm.

It is always easy to find my companions. We are the only ones on the bay at this hour. Hank is on the far shore loading his boat, Mayo is preparing to go fishing, Stevie has just rounded Sprague's Point, and Richard is puttering up the river. Soon we have all collected in midbay.

"Let's go catch some stripers."

"Nah, tide's all wrong. We'll never catch any."

"Menhaden are in. Saw some blues chasing 'em."

"Yeah, but the tide's goin' out. Let's go look for blue crabs."

Logic wins the day. Freshwater flowing out of Lonnie's River clears the water so you can see the blue crabs lurking beneath large bright-green mats of sea lettuce.

We cut our engines and let the outgoing tide carry us downstream. Beautiful red and yellow sponges litter the deep channel in the center of the river. We have trained our eyes to ignore the greenish-gray camouflage of the crabs, to look instead for the crimson edge of a claw or the faint outline of a shell. The trick is to plunge your net in front of the crab so it will swim into the net while scuttling toward the safety of the overhanging bank.

Sometimes we are lucky and spot a large male crab sculling across the surface. It's easy to scoop him up before he dives toward the bottom. Once we have the crab in the bucket, there is a mad scramble to slam a board on top. You have to be quick. More than once we have gone home with fingers lacerated by the lightning-quick slashes of a blue-crab claw.

After an hour the tide has slowed and the water has turned

turbid once again. The wind is rustling the silvery undersides of the poplars on Kent's Point. A school of menhaden is circling through the bay, filtering plankton from these productive waters. Occasionally the school erupts into a thousand fusiform bodies. Bluefish are hunting them from below. We hook up some minnows and soon have half a dozen snappers.

The snappers whet our appetite for bigger game. We decide to head down the bay to search for striped bass. But this is a dangerous time. The Namequoit Sailing Association is in session, and most of our parents are members. If we are not careful, they will shanghai us into crewing for one of their adult races—a fate worse than Latin or dancing school. The only way such an experience can prove to be in the least bit educational is if you happen to end up in the boat of one of the more articulate adults. Then you might pick up a particularly salty swearword or learn a new term for a risqué piece of human anatomy. Sometimes we hear hints about mysterious activities among consenting adults. These are reported to our group, and we spend long hours discussing and speculating about the exact nature and details of the behaviors so obliquely referred to.

But today we manage to elude our parents, hug the far shore, and scoot down the bay without losing any of our members. Our destination is the flats in Crooked Channel. The incoming tide sweeps silversides over the flats and into deep holes where schoolies lie in ambush. The excited calls of terns draw us to the spot. The silversides are caught in a vicious cross fire. Feisty terns dive from above, voracious bass slash from below. The water is full of fish scales glittering toward the bottom, the air smells of fish oil and plankton.

We fish furiously until the tide slows and drifting rafts of eelgrass start to foul our lures. We decide to head to the outer beach. There is still time to have some lunch, take a dip in the icy cold Atlantic, and dig some of the fat white steamers that thrive in the glistening clean sand on the inside of the outer beach.

By this time the tide is dead low, and we have to drag our boats across the flats. The incoming tide does the rest; it pushes us deep into the marsh behind the islands. But now we have enough water for our favorite game, "ambush."

I'm in the first team of two boats. While the second team counts, we roar into the labyrinth of creeks and are swallowed up by thick banks of marsh grass that tower over our heads. We find a likely spot, and back our two boats into small creeks on either side of the main channel. We feel like Athenian sailors waiting for our enemies' triremes to blunder into the Strait of Messina. But I suppose we are really more like the Vietcong.

The second team of boats putter slowly past our positions.

"Shh, not now, not now. Don't move."

They continue to where the creek narrows and starts to turn.

"They're trapped! They're trapped! Now, get 'em!"

"Fire! Fire!"

"Hurry before they turn!"

With all oars splashing, we blast out of the side creek with the sun behind us. It is a glorious victory, but the conquered refuse to submit. We have to settle the dispute with a long, drawn-out, quite spectacular water fight that leaves us drenched, spent, and almost swamped. We call a truce and pole through the Horseshoe to Sampson's Island, where we can clean and bail our boats.

But now our play becomes less innocent. Someone spots a horseshoe crab. Instantly, we are transformed into a bunch of crazed Ahabs launching our oars into the defenseless creatures. I'm afraid I still remember the gratifying crunch that the shells make when they receive a direct hit.

Such blood lust is sanctioned on Cape Cod. We have all been raised to believe that the only good crab is a dead crab. Most towns still have a bounty on horseshoe crabs because they are considered to be predators on shellfish. Children are encouraged to catch any horseshoe crab they can find, wrench off its tail, and throw it above the high tide mark. The tails can be turned in to the shellfish warden for a penny a tail or so.

Our family even has a dog who has picked up the habit. Jake now spends long hours wading through the shallows, pawing the bottom in search of horseshoe crabs. When he finds one, he plunges his head under water, grabs the crab by its tail, and hauls it onto the beach. As soon as the crab rights itself and starts to crawl toward the water, Jake digs a hole in front of it. When the hapless crab tumbles into the hole, Jake covers him with sand and starts searching for a second crab.

Jake has managed to teach two succeeding generations of dogs to follow his example. But the behavior has degenerated. The puppy only carries the crabs onto the beach and barks at them all day, much to Jake's disgust.

But now the sun is setting, and the tide has filled the bay once again. I must motor home in time to clean the fish on the bridge above the creek. It is slow work. I have to check the stomach contents of each fish and feed my favorite crabs in the creek I'm standing in. Impromptu anatomy lessons keep getting in the way. My hunter-gatherer instincts are already starting to evolve into more systematic study.

In fact I have a theory that such hunter-gatherer behavior is necessary to the true development of biological curiosity. If so, I fear future scientists and doctors will miss out. They will never have the chance to discover nature for themselves. They will never have a chance to dissect a real organism.

Of course this is probably a hopelessly out-of-date, politically incorrect theory. In an era of hockey camps and tennis lessons it seems downright primeval to suggest that young people can actually keep themselves occupied without television, the Internet, or adult supervision. But I know we did, and it set the course of our lives forever. So I will stick to my theory until somebody proves it wrong.

In the meantime, whenever I elect to have minor surgery, I will continue to go to my old friend Mayo Johnson. He is chief of surgery at a major hospital outside Boston. I know he has good technique; I watched him perfect it filleting flounder.

Not far away a different kind of doctor was also perfecting his technique.

CHAPTER 2

Carl Shuster

[July 14, 1953]

THE ADMIRAL'S HOUSE overlooks Vineyard Sound and the Woods Hole Coast Guard Station. It is surrounded by towering beech trees, whose copper-colored leaves rustle in the winds that blow faithfully off the sparkling waters of the sound. The beech trees were planted by Joseph Story Fay, the first Bostonian to make Woods Hole his summer home. He planted the trees throughout the village in the late 1800s, and now they gave it an air of elegance missing from the scrubbier lowlands of the outer Cape.

Carl Shuster loved the beech trees. People had so much more faith in the future in those days. Mr. Fay had planted the trees knowing that they would only come into their full glory a hundred years after his death. Carl wondered if he would ever make such a long-lasting improvement to a place he cherished.

But at the moment Carl had more immediate concerns. He was trying to finish packing the car before breakfast. He could already smell the eggs and bacon Helen was cooking for the twins. The double blessing of two sets of twin boys had taken the Shusters by surprise. Money was tight, but they couldn't ask for a better life. The Woods Hole Oceanographic Institution had given the Shusters the use of the Admiral's House for the summer. All they had to do in return was mow the lawn and act as houseparents to the four students who shared the other side of the mansion. It was not a difficult assignment; the lawn rolled down gracefully to the water, and the young men were all hardworking graduate students.

Eventually the old Crosley was loaded. It was supposed to be a station wagon, but Carl calculated it was smaller than the Volkswagens that Hitler had designed for the Volks of Germany.

The war was little more than a bad memory by 1953. Woods Hole had returned to its former civilian footing. Scientists no longer practiced antisubmarine warfare in the pens behind Naushon Island. Carl's mentor, Dr. Redfield, had spent the war studying hemocyanin, the copper-based blood of horseshoe crabs. It had been Dr. Redfield's suggestion to study the horseshoe crabs of Pleasant Bay.

Carl always looked forward to the drive from Woods Hole to Pleasant Bay. The only towns of any size were Falmouth and Hyannis. But they slipped by quickly, and from then on it was only rural roads and tree-lined farmland. Carl kissed Helen and the kids good-bye. They would spend the day exploring the shallow waters of Woods Hole while Carl did the same on Pleasant Bay.

When he reached Orleans, Carl contacted the shellfish warden, Elmer Darling.

"Mornin' Constable. Looks like another beauty, doesn't it?"

"Sure does. Got the Spurges' skiff all lined up for ya. Just tie 'er back up when you're through with 'er."

Carl loaded his measuring board, logbooks, pencils, and buckets into the skiff and headed out slowly into the bay. The tide was already flooding, so he decided to skirt around Hog Island and head directly to the spawning beaches.

Carl already had a reputation among Woods Hole scientists as being a bit of a bucket biologist. The lab scientists used the phrase as a mild term of derision, but to Carl it was an expression of deepest respect. He loved to be outdoors and possessed an intuitive sense of nature. As a boy in New Jersey he had raised chickens through the 4-H program and remembered feeding the hens crushed horseshoe crabs to thicken their eggshells.

But now Carl's sense of nature was fully engaged. A windrow of crab shells lined the shores of Hog Island. Carl tucked the observation into the back of his mind and continued toward the outer beach. As a teacher he always told his students, "Look for the anomalous thing. That's what's important. In nature, it's the exception that proves the rule."

Soon the tide was in full flood, and horseshoe crabs were using the currents to push them forward. They would crawl on the tips of their claws half-walking, half-surfing, scuttling toward the beach. Carl jumped out of the skiff and started wading through the shallow water. It was the time-honored method of catching horseshoe crabs. He would stoop, scoop up a crab, measure it on the measuring board, record the data, and put the crab back in the water, all in one easy motion. Sometimes he saw a blue crab or a puffer fish hovering behind a female horseshoe crab, as if it was expecting to eat her eggs. He recorded each observation faithfully in his field notes. It was pleasant, easy work. The water was crystal clear, and the warm sun shone brightly on his tall gangly frame.

Carl always preferred wading without shoes so his toes could feel for quahogs buried in the sand. These were placed directly into a wire basket, destined not so much for science but for the culinary arts. He and Helen would shuck the clams for chowder. More than once this free source of protein had helped them make it through the winter. It was just one of the many perks of being a field biologist.

But then something caught Carl's eye. One of the male crabs had a piece of an old claw attached to its clasper. Horseshoe crabs normally molt every year. But during the tenth year the forward claw of the male crab metamorphoses to become a fist-shaped clasper. The male uses this clasper to grasp a female during mating. Normally the old claw drops off as soon as the male mates with a female. But this male still had a piece of atrophied pedipalp attached to its clasper.

"So, you're still a little virgin, aren't you little guy?"

The male flailed his legs quietly in Carl's gentle hands. Most people would have regarded this extra piece of claw simply as an interesting curiosity, if they noticed it at all. But the observation niggled at the back of Carl's brain. What could this tell him about the horseshoe crab's life?

Carl turned the observation over in his mind as he continued to measure the crabs. By the end of the day he had measured 20 more virgin males and 195 nonvirgin males. Now he had a plan. He remembered the dead crabs on Hog Island and returned to measure them as well. He was shaking with excitement: just by measuring these crabs, Carl felt he could nail down an answer to a question that had stumped experts for years.

Carl finished his work and drove back to Woods Hole in a distracted state of mind. After supper he walked down to the Marine Biological Laboratory to go over his data. Like so many scientists before him, Carl loved the MBL library. It stayed open twenty-four hours a day, every day of the year except Christmas. Carl had searched its stacks for beautiful old monographs and folios on horseshoe crabs that could be found nowhere else on earth. Many dated back to the eighteenth and nineteenth centuries.

Carl always chuckled when he entered the library. You had to pass under a sign that bore Louis Agassiz's stern admonition: "Study Nature, Not Books." It was not a bad saying to have hanging over the front door of the world's largest marine library. Yet Carl knew the importance of good libaries—and good teaching. He could trace his intellectual awakening to similar words passed down from Louis Agassiz to William Keith Brooks to Julius Nelson, father of Carl's mentor, Thurlow C. Nelson.

It was Dr. Nelson who had first handed Carl a jar of horseshoe crab eggs with the laconic advice, "Study these!" It was an age-old teachers' trick: ignite a spark of curiosity, fan it slightly, stand back and get out of the way. With luck you would witness the blossoming of a chain reaction of learning inspiring to behold. Carl smiled at the memory of his mentor's enigmatic advice. Here he was, ready to spend the night pondering the meaning of horseshoe crab measurements. He suspected Professor Nelson had it figured that way all along.

But it was time to get down to business. Carl set out his field notes and started to work. In three days he had measured 115 female and 288 male crabs. But it was the male crabs that interested him the most. He separated them into the virgin males with the attached pedipalps, the middle-aged males he had found crawling on the bottom, and the old males with worn shells he had found lying dead on the beach. Then he showed the numbers to a colleague. This was something of a tradition at the Marine Biological Laboratory. Late at night you could be led astray by your own enthusiasms. It was always best to share your data with a skeptical colleague.

"OK, now look at this. Here are the virgin males. The average width of their shells is 166 millimeters. Now look at the middle-aged males. The average width of their shells is 167 millimeters. And over here, the average width of the old males is only 168 millimeters."

"They're all the same, so what?"

"Don't you see? These old crabs are ten years older than the virgin crabs, but they're still the same size. That means they stop growing and shedding their shells after they reach sexual maturity."

"What about the females?"

"That's the point. The females keep growing and keep shedding their shells. Sometimes they get to be almost twice the size of the males."

"That doesn't make any sense. You would think the males would keep growing so they could compete with other males. That's how most animals do it."

"But the males don't need size to compete. They just cluster around the females and release their sperm. In fact a smaller male might even have an advantage because he could crowd in closer to a female to inseminate her eggs."

"But why do the females keep growing?

"Don't you see? The bigger the female is, the more eggs she can produce."

"Good God, Carl, you know I think you might be onto something."

"Thanks, but—oh my gosh, look at the time. It's already one o'clock. I've got to get up early tomorrow morning."

"You have it made! In the field all day, in the library all night."

"Hey, it sure beats being one of you workbench biologists."

"Get outta here."

In future years the Shusters would buy a bigger car and head farther into the field. The twins had grown older, so Helen could join Carl in his travels. They attached a plywood box to the top of the car to hold horseshoe crabs and set out to explore the East Coast. They started in Maine and drove all the way to Florida, stopping in estuaries to count and measure horseshoe crabs.

In northern New England the crabs were tiny. From Cape Cod south they started to get bigger. In Delaware Bay they were huge and swarmed onto the beaches by the hundreds of thousands to lay their eggs. After Georgia the size of the crabs started to decline again. By the time the Shusters reached the white limestone sands of Florida, the cream-colored horseshoe crabs were again the size of the crabs in Maine. They remained that way down through the Florida Keys, into the Gulf of Mexico, and on down to the Yucatán peninsula. Carl proved that temperature was what determines a horseshoe crab's size and that they are truly a temperate-zone species. But a good part of the experience was traveling the open road, exploring the nation's estuaries, staying with friends, and visiting marine labs up and down the East Coast.

Carl thanked his mentors and teachers for introducing him to such wonderful animals that crawl right up to your feet and display their most revealing, intimate behaviors. In later life Carl would counsel his students to select their animals with equal care. "Whales and gorillas are fascinating, but you don't want to spend your whole life scrounging for grants to fund costly research vessels or expensive field trips to Africa. Select a nice interesting animal that lives close to home and is easy to get to. You can build a career around such an animal, and no one can snatch it away from you!"

Carl did indeed build a career around horseshoe crabs, but he never forgot those wonderful days on Pleasant Bay when he was making his first discoveries. Little did he know how important they would become fifty years later. Little did he know how much of a lasting improvement he would make to the East Coast, which he grew to know and love.

CHAPTER 3

First Lessons

[Boston Children's Hospital, 1956]

"MRS. SARGENT, your son is alright but I'm afraid there's been some brain damage. He's talking some nonsense about worms, crabs, and bugs."

"Oh, that's alright. All he ever does is talk about worms, crabs, and bugs."

During my early years I was obsessed with three things: horseshoe crabs, pill bugs, and the creek that emptied the marsh behind our house. All were readily available within a short walk. But the closest were the pill bugs that lived beneath the pine needles at the foot of our bulkhead. Every morning I would sit on the bulkhead, soaking in the sun, munching my tomatoes, and playing with pill bugs. All I had to do was brush aside the thin cover of pine needles and there would be friendly little congregations of them, hiding under the litter or clustering along cracks in the concrete where moisture accumulated.

As soon as I disturbed the pill bugs, they would race back toward the cool dark safety of the litter. But they would do this in a curious manner. First, one would find a leaf, slow down, and stop. Then another would bump into the first, slow down, and stop. Finally all the bugs would be reassembled into their cozy little assemblages. This behavior goes by the technical term thigmokinesis, roughly translated as "running into things."

But the most fun was to pick up the bugs and let them roll around in the palm of my hand. As soon as one of the creatures was disturbed it would curl up into a shiny gray sphere, so perfect it looked like a man-made ball bearing. This was the bugs' impregnable defense. Wolf spiders would palpitate the spheres for minutes, looking for a chink in their armor. Shrews could only roll the spheres around ineffectually before leaving in a huff.

Sometimes I would investigate the pill bugs with my hand lens. They had a head, fourteen sturdy little legs, and seven shiny gray body parts. Even then, I knew the critters were neither spiders with eight legs nor insects with six. But they did bear a striking resemblance to tiny white- and gray-speckled creatures who would also roll into a ball when we found them in salt water while digging for quahogs.

In fact, whenever I brushed aside the pine needles, I had the uncanny feeling that I was gazing into an ancient Paleozoic sea. The pill bugs looked as if they were in an animated museum diorama, like one I had seen at Harvard's Museum of Comparative Zoology. It had a slew of trilobites scuttling beneath a Precambrian reef to escape the groping tentacles of an ancient squid.

Little did I know, I was almost on target. Pill bugs are distantly related to trilobites, and their closest surviving cousins are isopods like the ones we disturbed while digging quahogs in the bay. You have probably seen gribbles. They are those little crustacea that swarm over docks and pilings. They riddle the wood with half-inch-deep holes and can destroy an unprotected dock in less than a decade. If you frequent nude beaches, you have probably also discovered another pill bug relative, *Idotea baltica*. These beautiful, green- and white-speckled, three-quarter-inch isopods live in clumps of eelgrass. If disturbed, the first thing they do is swim straight into the nearest dark clump of stringy grasslike material they can find. If you happen to be swimming nearby you might notice one making a beeline for your pubic hair. I can speak only for myself, however. If you are at a nude beach, your attention might be drawn elsewhere.

Like their marine cousins, pill bugs still breathe with gills. This is the reason you find them clustered around areas of moisture. In fact pill bugs seem to be suspended in midevolution, living on land but with a body designed for water. In their former lives, pill bugs evolved short sturdy legs for scurrying across the ocean bottom. They developed their flattened bodies to contend with water pressure, and they developed water-filled pouches to protect their larval offspring—all advantageous on land as well.

In retrospect, I find it interesting that I became so fascinated with pill bugs, a land animal whose behavior is so closely related to that of horseshoe crabs. You can think of a horseshoe crab as living halfway between the marine world of a trilobite and the terrestrial world of a pill bug. Horseshoe crabs come onto land only once a year to lay their eggs. If disturbed while spawning, they curl up much like pill bugs to protect themselves. As long as they keep their gills moist, they can survive for several days in this defensive curled-up posture. The blood of both animals is based on copper, not iron. Some scientists think that their common ancestor evolved this blue copper-based blood when copper was more available in the oceans than iron. This is still a problem for pill bugs. They have to eat their own feces to recycle copper back into their blood.

But feces eating has left pill bugs open to another threat, parasitism. When pill bugs eat the feces of songbirds like starlings, they ingest the eggs of a parasitic worm. Relatives of this worm still live in the ocean, where they parasitize arthropods and fish. However, this worm, the spiny-headed acanthocephalan, evidently followed pill bugs onto land, where it now parasitizes pill bugs and birds, not arthropods and fish.

The pill bugs pick up the worms by eating bird turds. That's pretty straightforward. But how does the worm manage to get from a pill bug into a songbird for its adult stage? Remember, pill bugs are timid nocturnal creatures, but songbirds feed by day. How do the worms do it? Well, the worm has evolved an ingenious method. It alters the pill bug's personality. That's right. Instead of being shy, retiring nightime creatures, pill bugs infected with acanthocephalans become bold, extroverted risk takers that think nothing of streaking across your front yard in broad daylight. Of course the inevitable happens. Starlings swoop down and gobble up the raging extroverts.

Does this sound familiar? Pill bugs changing their personalities? Well, it should. Acanthocephalans change their host's behavior by increasing the amount of serotonin coursing through pill bugs' little brains. And we thought we were so clever inventing Prozac. The spiny-headed acanthocephalans discovered the serotonin reuptake mechanism hundreds of millions of years before us. Interesting, though, that they used it as a weapon rather than a mood enhancer.

I learned from books the finer points about pill bugs' mood swings. But it was in the creek where I learned my first lesson in conservation. Every day I would spend time observing the lives of minnows, shrimp, and crabs. I would feed the crabs and knew each by name, location, and behavior. The small male blue crab lived under the bridge, the large female lived in the bank downstream, the timid blue with one claw lived upstream and would wait for the others to eat first. I had an unwritten law that occasionally I could harvest a single male blue crab, but I knew it would be missed, and I knew that it would take several weeks for another crab to reoccupy its empty niche.

Then one day it happened. A group of roving teenagers sauntered up the shore armed with nets and buckets. It was obvious they meant business. Initially I was cordial, offering to show them where a large male blue crab could be caught. But that was not enough. The band proceeded to march, two by two, up either side of the creek, flushing the hapless crabs between them. The crabs didn't have a chance. After they were netted, the boys would smash the crabs' shells to immobilize them while the little visigoths continued with their slaughter. By the end of the day two dozen crabs lay dead or dying. "George," "Harry," "One Claw," "Crabette"—all lay with guts exposed, bleeding into the marsh grass.

The boys had killed every crab in the creek. It never recovered all summer. I trudged back to the house feeling helpless, sad, and violated.

That night I developed a headache, felt nauseous, and had diarrhea. I had to get up several times to vomit and started feeling weak, dehydrated, and delirious. By the following morning I had slipped into a coma.

Joe Gallagher was summoned. He was on old-fashioned country doctor known more for his compassion than for his hygiene. He always had a fat cigar clamped firmly between thick voluptuous lips. A constant rain of ashes fell onto his considerable paunch. Often they dropped into the water he was boiling to sterilize his needles.

Dr. Gallagher returned for several days, but my fever never broke. I'm told he had tears in his eyes when he confronted my parents: "I'm terribly sorry, but there's nothing more I can do. You better get your son up to the Children's Hospital as fast as you can." I think I remember the ambulance, the sirens, and my parents driving behind me as we tore up Route 6 toward Boston. But perhaps I'm mistaken.

At Children's the doctors ran tests, took electroencephalograms, introduced antibiotics, and waited.

One kindly nurse became particularly interested in my case. Then one day she met my mother at the door. "Oh, I'm so glad you're here. Your son woke up this morning but I'm afraid there has been some brain damage. He's talking some nonsense about worms, crabs, and bugs."

My mother gave a great sigh of relief.

"Thank God! No, he's perfectly alright. All he ever does is talk some nonsense about worms, crabs, and bugs!"

CHAPTER 4

At an Ancient Orgy

[June 21, 1957]

THE MOON IS FULL. There is no wind. There are no sounds save for the quiet lapping of the bay against the shore. I have taken the skiff to the far backside of Nauset in order to witness one of nature's great spectacles, the annual spawning of horseshoe crabs. On June 21, the twin forces of the sun and moon have conspired to make this the first day of summer, the longest day of the year, the highest tide of the season. Now these invisible forces are drawing the tide inexorably higher.

I slip into the moonlit waters. It is like diving back in time. The dark form of a large female horseshoe crab looms into the silvery luminescence. She is preparing to reenact a ritual that has persisted for over 450 million years. Her ancestors lumbered out of similar waters to lay their eggs before there were birds, fish, or dinosaurs to watch. The only things living on land at the time were mosses, ferns, and occasional invertebrates like dragonflies with four-foot wingspans.

But the female is not alone. A smaller male crab crawls toward her and executes a brief circling dance before clasping her shell with his specially modified mating claws. Tonight, his tenacity will pay off.

As the couple make their way toward shore they must navigate through a stag line of eager male suitors. The males clamber over each other in their eagerness to clasp onto her carapace. By the time she reaches shore the female is dragging two more ardent males.

The tide slows. Rivulets of water pause in their steady advance shoreward. The moonlight now glistens on the shells of thousands of crabs that have crawled onto the beach. Each female is surrounded by thirty or forty lascivious males. I can hear the quiet scraping and scratching of the crabs' shells as they climb over each other in their eagerness to lay or fertilize the eggs.

The female digs into the sand, deposits several thousand eggs, then drags her mate over the eggs to fertilize them. But her mate is not alone. The other males swirl around the pair in the hope that their sperm will outcompete the mate's. The water is now replete with eggs, sperm, and pheromones, chemical aphrodisiacs the female has released to attract the males.

Within thirty minutes it is mostly over. The tide has turned, and hundreds of spent crabs crawl slowly back into the bay. Those that are too exhausted to make it dig into the sediments to await the next tide. To fail to do so is to be stranded to a certain death in the heat of the morning sun. In the end, all that remains are thousands of fertilized eggs, their cells already dividing unseen beneath the sand.

The following day I return to the beach with Charlie Wheeler and my father. Dr. Wheeler is a biologist who is studying the role of horseshoe crabs as shellfish predators. He is an old hand; he collected horseshoe crabs for Dr. Redfield, a renowned Woods Hole scientist who studied horseshoe crab blood during the war as part of the war effort.

Dr. Wheeler strokes his beard and scrutinizes the beach. "Hmm. See that little depression over there? That's probably where they were mating last night. Looks like a likely spot."

"Yup, that's about where I saw 'em."

Charlie drives a short flat-ended garden spade into the ground.

"Bingo!"

Sitting beneath six inches of sand are three thousand perfectly formed green eggs the size of a pinhead.

"What'd I tell you? Wash these through the sieve, and let's get some more."

Farther up the beach we discover another clutch of eggs.

"Hey look at this. Now this is a real find."

"What is it?"

"These are also horseshoe crab eggs, but they're about ready to hatch."

"They look like cellophane."

"Yeah, they do. These are what we call bubble eggs. They were laid during the full-moon high tide a month ago. But about two weeks ago their green outer shell broke open, and this larger transparent egg bulged out. There's only a few hundred eggs left. I suspect most of the clutch hatched out when the high tide ruptured their shells last night. The others will probably hatch out during today's high tide."

"That's slick."

"It sure is. Here, put 'em into this container. We'll take them back and look at 'em under your microscope."

Back at the house we pipetted an egg into the shallow well of a slide, then slipped it under my microscope. My parents had given me the much coveted microscope as a gift to celebrate my recovery from encephalitis. To my young years the microscope was well worth the disease. I had been hounding my parents for a decent microscope for several summers.

Careful not to crush the cover slip, I brought the egg into focus. It was like being transported back to the very beginnings of life on earth. Inside the egg I could see a half-inch, grayish-green creature twitching, swimming, revolving slowly within its transparent sphere. It already had two prominent eyes and looked vaguely like a horseshoe crab, but it was rounder and lacked a tail.

"It's so prehistoric!"

"Indeed. This is probably what trilobites looked like half a billion years ago. In fact this is what's called the trilobite larva. Soon it will metamorphose into a proper little horseshoe crab."

"It's pretty big right now."

"Yup, it's already shed its shell twice. Look on the bottom of the egg. See those little pieces of tissue that look like discarded cellophane wrappers? Those are the exuviae of its earlier larval forms."

I was flattered that Dr. Wheeler used the proper technical terms. It felt like I had passed some kind of rite of passage, was coming of scientific age. Perhaps he recognized that my hunter-gatherer instincts were developing into a more science-like perspective.

But that scientific perspective did not protect me from making a grave mistake in later life. During those early summers I spent many hours diving with horseshoe crabs both by day and by night. In those dives I couldn't help but notice that during the day horseshoe crabs seemed almost totally blind. They would often blunder right into my bare feet. Yet at night they seemed far more perceptive. The males would crawl unerringly toward the females. It seemed likely that the females were releasing something to attract the males.

I had slipped into the wrong camp of the great pheromone debate. In the 1960s scientists discovered pheromones, chemical aphrodisiacs that animals use to attract their mates. Lobsters use them underwater, and the lowly cecropia moth uses pheromones to attract mates from as far away as three miles. When I heard about these strange attractors, it made sense to think that horseshoe crabs might use them as well.

It was only in 1986 that I realized my mistake. I was sitting in the kitchen discussing the work of Dr. H. Keffer Hartline, who won the Nobel prize for his research on the optic nerve of horseshoe crabs. Suddenly my mother, who had by now become used to such conversations, piped up: "Wasn't there a Dr. Hartline who spent the summer in the Bentons' old barn? He was studying something about horseshoe crabs. Remember Fay [my older sister, famous for her frequent and intense love affairs] had a crush on Danny Hartline."

"Good God! You mean Dr. H. Keffer Hartline was living right next door and no one ever told me about it?"

"I think he was some kind of doctor."

"Some kind of doctor? He won the Nobel prize for God's sake. Why wasn't I told?"

"Now there's no need to take the Lord's name in vain, dear."

Suddenly the memories of that fateful year came flooding back. All we had heard about for the better part of the summer was this dashing Danny Hartline. Evidently he was churning up my sister's already hyperactive hormones during long midnight sailing trysts. There she was, hanging around the house, trailing a scent of undoubtedly powerful pheromones, totally oblivious to the pioneering research on horseshoe crabs going on right under her nose. I was appalled at my sister's abysmal sense of priorities. Of course, not only was Dr. Hartline doing some kind of research on horseshoe crabs, but he was well on his way to unraveling the mechanism that underlies vision itself. Ever since that time I have made it a point to observe my sister's rich and varied love life. It remains a fascinating subject, but alas, not the subject of this book.

So how did the pheromone argument get resolved? One of Dr. Hartline's graduate students, Bob Barlow, put cameras on male horseshoe crabs and discovered they would make a beeline toward plaster models of horseshoe crabs that couldn't possibly be giving off pheromones. The horseshoe crabs were using their compound eyes to see their mates.

But something else was also going on. Dr. Barlow discovered that when ultraviolet light strikes a horseshoe crab's simple eye it changes the structure of the crab's compound eyes so they see better at night. When is there more ultraviolet light? During the full moon. When do horseshoe crabs need to see best? When they mate. So who needs pheromones if you have a full moon? Of course, Dr. Barlow could have saved himself a lot of hard work if only he had listened to that famous neurobehaviorist Dean Martin when he sang, “When the moon hits your eye like a big pizza pie, that's amore!”

But not far away another scientist was making making a curious observation about horseshoe crab blood, an observation that would profoundly affect the lives of humans and crabs.

PART II

Commercialization

CHAPTER 5

The Conversation

[Woods Hole, 1962]

APPROACHING 90, Betsy Bang is an attractive, sprightly woman with an elfin haircut and an impish sense of humor. "This is Marmaduke," she says, indicating a photograph of a silverbacked mountain gorilla. "Marmaduke introduced me to Frederik. There I was, perched on the corpse of this 700-pound sweet-smelling gorilla in the Johns Hopkins dissection room. In pops this tall, bespectacled, bearded medical student.

" 'What are you doing?' he asks.

" 'Why, I'm drawing Marmaduke's musculature. What are you doing?'

" 'Oh, I just finished an autopsy next door. But why are you making drawings of a gorilla?'

"I told him it was for a book on gorilla anatomy, as if it was the most natural thing in the world.

"I can't quite remember how the romance proceeded after that, but our lives were never the same again. We bought this house in 1950. That was the year that Stephen Kuffler invited Frederik to visit the MBL. You remember Steve—what a charmer. He wrote the book on neurophysiology and should have won the Nobel prize. Anyway, Frederik stopped by to see the lab and was just amazed. He had no idea that you could learn so much by working with simple marine animals. We bought this house immediately. Frederik knew Woods Hole would somehow be the focus of his career. So you can see, it was all Stephen Kuffler's fault.

"During those early years we had a wonderful routine. Every morning Frederik would ride his bike down to the lab while I fed the kids and drove them to the School of Science. What a wonderful program that is, harder to get into than MIT or Harvard. Later I would join Frederik in the lab, where I was working on the sense of smell in birds. You know, Audubon had it all wrong when he said birds had no sense of smell. That put the field back fifty years.

"That first year Frederik worked with many animals. But the first time he injected some bacteria into a horseshoe crab, he knew he had discovered his research animal. You see, it did the most amazing thing. It simply sludged up and died. Later he found out that the same thing had happened when Noguchi injected the blood cells of a horse into a horseshoe crab way back in 1903.

"Well, most people would have simply thrown the crab away and picked up another. But Frederik realized something special was going on. Here you had this ancient creature that didn't have an immune system—it lacked both T cells or B cells—but it seemed to be mounting some kind of defense. Frederik knew he had stumbled onto a simple model for the human immune system, but he couldn't get anyone to listen for years. The medical establishment is pretty conservative, you know.

"Finally we decided that academics think better when they are on vacation, so we invited Lockard Conley to stop by Woods Hole on his way to his vacation home on Martha's Vineyard. Of course we had known Lockard for years. He was head of the Hematology Department at Hopkins Medical School. So, I remember Frederik biking down to the village early one morning in 1962 to get his laboratory all set up. When Lockard arrived, the horseshoe crab was prepared. The conversation went something like this:

" 'OK, now we know how the human immune system works. If I get a wound, we have a whole array of cells and antibodies that fight the incoming infection. Some label the antigens, others ingest the invaders, some even release chemicals that go to the brain's hypothalamus and trigger it to raise our body's temperature.'

" 'Sure, that's the classic Schwarztmann reaction. Jack Levin's been working on it in rabbits.'

" 'OK. Now here's this animal that you would think lacks an immune system. But it lives in shallow waters that also teem with bacteria. Now watch what happens when I inject this crab with some vibrio bacteria. Here, look at some of the crab's blood under the microscope.'

" 'Why, it's turning a bright cobalt blue.'

" 'Right, and coagulating. Now, if you inject too much bacteria, the crab gets intravascular clotting and dies. But if you only inject a small amount of bacteria into the crab, it seems to be able to fight the infection and survive.'

" 'So it's like a simple immune system.'

" 'Yes. We've found that instead of having a whole array of immune cells, horseshoe crabs simply have these large amoebocyte cells. When a horseshoe crab receives a wound, their primitive cells swarm to the area, coagulate, and simply immobilize the bacteria.'

" 'So you're saying these amoebocytes act like platelets in human blood?'

" 'Exactly, but they use the clotting system to fight infection. So you see, it's a primitive immune system. Probably the oldest in the world.'

" 'Almost a precursor to our own.'

" 'Yes, and it evolved in an animal related to trilobites.'

" 'Elegant work, Frederik!'

" 'Now, if I could only work with a young hematologist . . .'

" 'OK, OK, you've convinced me. I'll see if can't get Jack Levin to work with you next summer in Woods Hole. I don't think he's ever seen a horseshoe crab!' "

The collaboration between Dr. Bang and Dr. Levin proved to be one of the most fruitful in the history of applied science. Over the course of several summers they learned how to stabilize the amoebocyte cells, lyse open their outer coatings, and make what is today called *Limulus* amoebocyte lysate, LAL, or simply lysate for short.

What Bang and Levin had done was to create a new way to test for bacteria. All you had to do place a small amount of lysate into a test tube, swirl in an equal amount of liquid to be tested, and watch. If a bright blue gel formed in the bottom of the test tube, you had bacterial endotoxin in your sample. It was as easy as using a grade school chemistry set. Word spread quickly. Here was a sensitive, fast, inexpensive new way to test for bacteria—an assay that could be done in a test tube rather than in a live rabbit.

Today, the first thing a researcher would do after developing an "in vitro" test to replace an "in vivo" test would be to hire a good lawyer and apply for a patent. But in 1968 biology was still in its prebiotech infancy. Dr. Bang and Dr. Levin simply published their paper with its explicit instructions for how to make stable amoebocyte lysate. It was arguably the most important discovery ever to come out of the field of marine biology, but publishing it would prove to be a costly mistake.

Mrs. Bang deserves the last piquant word. "They were really jerks. If Frederik had only applied for a patent, the MBL would have received millions of dollars in licensing fees by now. That would have funded a lot of research!"

CHAPTER 6

Bleeding the Crab

[1969–1974]

THE NEWS ABOUT LYSATE spread quickly through the research community. Dr. Jacob Fine from Harvard Medical School set up a lab at the MBL to see whether *Limulus* lysate could be used to diagnose diseases like spinal meningitis.

The notion was sound and had a distinguished history. In the 1880s Dr. Hans Christian Gram discovered a staining technique that could distinguish between two major families of bacteria. One group caused the stain to turn deep purple; the other produced no response. The first became known as Gram-positive, the second Gram-negative bacteria. When Alexander Fleming discovered penicillin in 1928, he noticed that the common bread mold killed Gram-positive bacteria more readily than Gram-negative bacteria. Scientists later discovered that it was the sugarlike outer coating of Gram-negative bacteria that blocked the Gram stain from turning purple.

But the coating of Gram-negative bacteria plays a far more sinister role. It is composed of endotoxins, poisons that cause the burning fevers associated with diseases like typhoid, spinal meningitis, gonorrhea, and toxic shock syndrome. In hospital circles endotoxins are usually called pyrogens, or "burning bodies," because of the high fevers they cause.

Gram-negative bacteria are as ubiquitous as they are lethal. Not only can they live in shallow waters with horseshoe crabs, but they also thrive in the human intestine, where they normally do little harm. It is only when Gram-negative bacteria enter the human blood system that they become dangerous. That might happen to someone in a car accident. Trauma causes the membranes between a person's gut and blood system to break down; Gram-negative bacteria can now insinuate themselves into the blood vessels and multiply. There they release endotoxins that cause a raging fever. The accident victim is now in septic shock. If not immediately treated, he or she will be dead in twenty-four hours. The same thing happens during toxic shock syndrome. Gram-negative bacteria living on tampons get into a woman's blood system during menstruation. In spinal meningitis, Gram-negative bacteria infect the meninges, the lining of the brain stem. The first symptom is simply a stiff neck and fever, the second—death.

Rapid onset and lethality are the hallmarks of these diseases. A test that would take only fifteen minutes to perform could be a lifesaver. It would certainly beat the rabbit test, which takes four hours to perform—after you've located a willing rabbit. That was Dr. Fine's notion when he set up his lab at MBL. Unfortunately it didn't work. Doctors could use *Limulus* lysate to diagnose whether they had a Gram-negative bacteria, but it couldn't tell them which antibiotic to give their patient.

But not far from Dr. Fine's lab another scientist was interested in using lysate for an altogether different purpose.

"Freddie, I just heard Jacob give a talk about his work with *Limulus* lysate. Do you think we could use it to test our deepwater samples for bacteria?"

"Might be worth a try."

Stanley Watson drew a long drag on his latest cigarette and placed it on the overflowing ashtray beside his papers. He was a large vigorous man, impatient with the slow pace of research at the Woods Hole Oceanographic Institution. "Should only take a few weeks, at the most. Think you can get ahold of some horseshoe crabs?"

This was a standing joke between the oceanographer and Fredericka Valois, his assistant. The Valois family spanned both laboratories in Woods Hole. Freddie worked for Stanley at the Woods Hole Oceanographic Institution, and her husband headed up the collecting department at the MBL.

Freddie grabbed some buckets and headed for the door. Too late. The sound of a boat horn reverberated through the village. It was the *Gemma gemma* signaling the bridge tender. Traffic came to a halt as the drawbridge slowly rose. There in bold print was last night's grafitti, "Tourists Go Home!" It was something of a tradition, painting grafitti on the underside of the Woods Hole bridge. Locals could rib scientists, grad students could mock administrators, and everyone could dump on tourists without fear of attribution, or reprisal.

By the time the bridge descended, Freddie had reached the collecting department. It was housed in an old Cape Cod shack with gray weathered shingles. Inside, the sounds of water were everywhere. Water splashed into tanks, sloshed over sides, slopped onto floors, and gurgled down drains. A school of dogfish swam circles around their holding tank. Occasionally one would rise out of the water, lash its tail, and peer over the side before resuming its ceaseless circling. Toadfish glowered from beneath clamshells in a nearby tank.

A group of scientists lounged against the wall, telling stories and trading gossip. They were neurophysiologists, waiting for the *Gemma gemma* to deliver their live squid. As soon as the collecting boat docked, the scientists scooped the wriggling squid out of the hold, plopped them into buckets, and ran back to their labs. It was important to get the squid back quickly. They had to be dissected while their nerves were still alive and healthy.

A senior scientist caught a lobster and threatened the recalcitrant crustacean: "You better behave or you'll be a clambake after tonight's demonstration." A high school student buttonholed a Nobel prizewinner for tips on his upcoming science project.

John Valois presided over this informal bazaar with genial aplomb. His collectors had developed special techniques for catching each specimen and kept the locations of their finds secret so they could harvest them for years. The easiest animals to collect were horseshoe crabs. All you needed was a strong back and a small boat. This morning his collectors had brought back several hundred crabs from Pleasant Bay. Now the crabs clambered over each other or sat quietly in the long shallow running-water tanks that lined the sides of the building.

Freddie approached the tanks. "Good morning, John."

"Why, good morning, Freddie. What brings you here?"

"Stan wants some horseshoe crabs."

"Well, we got 'em. Take as many as you want."

Freddie selected several large females and put them in her bucket. "Think you'll have time for a sail tonight?"

"If I can ever get out of here. Someone has to take the trustees to the island for a clambake."

Back at the lab, Freddie propped up Bang and Levin's paper and followed directions.

> One—Fold an 8 to 12 inch horseshoe crab, *Limulus polyphemus*, along the hinge connecting the thoracic and abdominal segments.
>
> Two—Swab the membrane with 70% ethyl alcohol.
>
> Three—Insert a 20 gauge stiff needle into the pericardial sac surrounding the heart.

The 20-gauge needle was the same size needle a veterinarian might use to inoculate a horse. As soon as the needle punctured the crab's heart, bluish-gray blood spurted out of the wound and flowed down the stiff needle into a hundred-millimeter flask. On meeting the air, the copper-based blood oxygenated, frothing and foaming into a cobalt-blue supernatant.

> Four—Add N-ethyl maleimide to stop cellular respiration.

This stabilized the amoebocytes so Freddie could centrifuge the blood and lyse open the cells to make *Limulus* amoebocyte lysate.

A few weeks later, Dr. Watson had finished his tests. But he still had several liters of lysate left over, so he gave them back to John Valois to sell through the MBL's collecting department. Demand was immediate. Researchers from around the world started ordering lysate through the MBL. That ended the day John received a call from the comptroller.

"John, I see you've been selling a lot of animals. Your department is accounting for most of the money coming into the MBL."

"Well, actually most of it's coming in for lysate. We're selling it for $250 a pop, $15,000 a quart."

"Good God! That will have to stop, John. The MBL could lose its tax-exempt status if we keep bringing in that kind of money."

"Oh my goodness. I never thought of that. Of course it will stop. I'll send the lysate back to Stan right away."

But by this time Dr. Watson also realized that a lot of money could be made selling lysate. This notion fit in with his personal plans as well. He was growing increasingly impatient with his low salary as a research scientist.

Real Estate was the way to make money on Cape Cod. Several years before, Dr. Watson had incorporated a small, mom-and-pop real estate agency called the Associates of Cape Cod. But it was now clear he could make a lot more money selling lysate than condos. He already had the company, so he could simply start selling lysate under the name Associates of Cape Cod. But first, Dr. Watson had to protect his idea. Unlike Dr. Bang and Dr. Levin, Dr. Watson convinced the Woods Hole Oceanographic Institution to patent a slightly more sensitive way to make lysate. Then he asked the administration to invest in equipment. But the institution turned him down flat.

As it turned out, this also fit Dr. Watson's plans to a tee. He bought back the patent, set up a bleeding facility in the basement of his house, and started making calls to colleagues cajoling them into buying his product. But it soon became apparent that there was a much larger market than academic researchers. The Food and Drug Administration required that every major pharmaceutical company have large colonies of live rabbits so they could test their drugs for pyrogens. If the companies could replace the live rabbit test with the horseshoe crab test, they stood to save a lot of money. In short order Stanley Watson was able to land contracts with several major firms. Eli Lilly reportedly gave him half a million dollars in start-up funds. In exchange they would get a guaranteed supply of lysate and free consultation in perpetuity.

By 1974, Dr. Watson was doing so well that he was able to move the bleeding facilities out of his house and into a local plumbing supply company. He was still getting his horseshoe crabs from the MBL collecting department. That's where I came into the picture.

During the previous summer I had been working with a group of students studying the ecology of Pleasant Bay. We had noticed that the population of horseshoe crabs was dwindling. It was becoming more difficult to find the large female crabs preferred by the bleeders.

We held a meeting to decide how to deal with the matter. Most of our crew wanted to sabotage the collecting boats, but a small minority preferred to negotiate with the infidels. The debate lasted late into the night. Finally, after gallons of cheap wine had been consumed and effective arguments presented, the democratic process prevailed. It was decided that we would dispatch a small but persuasive delegation to Woods Hole. They returned with a deal. We agreed to collect crabs for bleeding and experiment with returning them back to Pleasant Bay.

We bought two-by-fours and chicken wire and built large pens in the shallows of Pleasant Bay. Then we started to collect the crabs and hold them in the pens. Every two weeks a truck from the MBL would pick up 250 crabs, deliver them to Falmouth where they would be bled, and return them to us so we could monitor their recovery. For this we were paid a dollar a crab.

It seemed like a sustainable fishery. But it would only work as long as the take was limited to a hundred crabs or so each week. But Dr. Watson had bigger plans for the Associates of Cape Cod.

In 1974, Dr. Watson decided to apply to the Food and Drug Administration for a license to produce lysate commercially. Today it would take several years and cost several million dollars to apply to the Food and Drug Administration. But in the early 1970s biomedicine was in its infancy. As Dr. Watson observed in later years, "I didn't know anything about the FDA, or anything about applying for one of their licenses. I just filled in the papers and sent them back."

In relatively short order Dr. Watson had his license. There was an explanation. The FDA wanted lysate for its own reasons.

CHAPTER 7

Crabs and Ponies

[Chincoteague Island, June 22, 1972]

FRANCES COOPER sat in the motel trailer transfixed by the images flickering on her television set. Hurricane Agnes had just swept across the Florida Panhandle, spewing tornadoes and killing fifteen people. Now the rogue storm was regrouping in the Atlantic and searching for a new place to land. Sitting in an aluminum trailer on Chincoteague Island was not Frances's idea of a safe place to be. Should they be on this remote island, whose only connection to the mainland was a narrow, flood-prone causeway?

Of course the kids were loving the excitement. They always enjoyed their annual trek to Chincoteague. The Coopers had been coming to the island ever since Jim had finished his graduate work with Jack Levin. His dissertation had shown that horseshoe crab blood was more sensitive than rabbits' at detecting endotoxins. Jim had started working at the Food and Drug Administration right after completing his graduate work at Johns Hopkins, but he hated his job. Fortunately, he saw a way out. Horseshoe crab blood would be his ticket to Chincoteague.

Everybody who used rabbits to test for endotoxins, it seemed, was now looking into lysate. In fact Mallinckrodt had approached Jim right after he finished his dissertation in 1969. Like all the major pharmaceutical firms, Mallinckrodt was required to have a colony of four hundred live rabbits to test its products. Every new batch of drug had to be injected into three rabbits and the rabbits monitored for four hours. If any of them developed a fever, the drug was contaminated. Mallinckrodt not only realized they could save a lot of money by getting rid of the rabbits; they realized they could make a lot of money by starting their own lysate facility. However, Mallinckrodt was headquartered in St. Louis, a thousand miles from the nearest horseshoe crab. Their solution was to hire Dr. Cooper to set up their lab on the East Coast.

Jim, however, had taken a desk job with the FDA. But he could escape if he could convince the FDA to collaborate with Mallinckrodt. It was not a difficult sell. The FDA also had their own colony of rabbits to test for endotoxins. They were planning to expand their colony, which would take up valuable space on their campus in Maryland. If the FDA could eliminate the proposed rabbit colony, they stood to save the $40 million needed to construct a new building on the site. It would be not only good science but a brilliant bureaucratic move.

Jim drove out to explore Chincoteague in 1971. There he found an old oyster shack that could be cleaned up and converted into a small lab. It quickly became a mecca for scientists interested in horseshoe crabs. One scientist was Carl Shuster, who came out to study their populations. Jim and Carl became fast friends. On weekends they would catch boatloads of flounder and fry them up for their families. Another visitor was Don Hockstein, Jim's colleague at the FDA.

So now Don Hockstein and his family were with the Coopers monitoring the progress of Hurricane Agnes. Their kids had spent the weekend catching crabs and keeping them as pets. (Jimmy Cooper had been the hero of his second grade class the year he had put a live horseshoe crab into his little red wagon and hauled it all the way to school for show-and-tell.) The kids couldn't wait for July, when the Chincoteague fire department would round up the ponies on Assateague Island and swim them across the narrow channel to Chincoteague to be auctioned off. But now the two families were doubly grateful for Assateague. The offshore island was protecting Chincoteague from the full force of Agnes.

Everyone on Chincoteague could remember the Ash Wednesday storm that had devastated the island in 1962. It had hit the day the new moon was pulling in the March spring tide. Thirty-foot waves had pounded Assateague Island, and a ten-foot storm surge had swept into Chincoteague Channel. It tore fishing boats off their moorings and flooded Main Street under six feet of water. Watermen had had to pole skiffs up Main Street and into the old Methodist Church, where water was lapping over the pews. For days the only communication between island mainland had been through a shortwave radio connecting the sheriff's office to a deputy's car.

Now here it was, ten years later, and it looked like another storm would repeat the damage. By the end of the day everyone was exhausted from yelling over the sound of the wind shaking the trailer. The Coopers fell into a fitful sleep wondering if they would wake to the sounds of popping rivets as the roof of their trailer flew away. The kids went to sleep still hoping the electricity would fail. Much of the East Coast, in fact, went to sleep wondering where Hurricane Agnes would come ashore.

The next morning the truth was revealed. The storm had miraculously bypassed Virginia and was on its way to a devastating landing in New York. But the wind was still raging, and the rain was still beating its incessant tattoo on the roof of the aluminum trailer.

"I wonder how the Boy Scouts are doing on Assateague?"

"Their tents must have collapsed. We'll probably see a sad little convoy of soaking children wending its way back across the causeway."

Little did the Coopers and their guests know that thirty people had died that night in Virginia alone. Thirty-foot waves had assaulted the fragile barrier islands. But all in all, Chincoteague had been fortunate. Other areas would not be so lucky. Pennsylvania suffered the worst natural disaster in its history. New York suffered a similar fate. A hundred people drowned, and $3 billion worth of property was damaged. Hurricane Agnes would go down as the East Coast's most damaging storm to date.

After eating a cold lunch, Jim and Don decided to check on the condition of the new lab, which was across the causeway. As soon as they unlatched the trailer door to leave, it burst inward, and the vehicle buckled with the impact. Both men had to pull on the door with all their might from the outside so Frances could latch it again from the inside.

Jim and Don inched their way across the flooded causeway. They were trying to get to NASA's rocket-launching facility on Wallops Island, where Jim had rented space in one of the old dormitories that housed technicians during NASA's early days. Now it housed Mallinckrodt's new horseshoe crab bleeding facility. To their amazement the lab had suffered only minor water damage. There was the rack Jim had built from lumber left over from a youth club Christmas crèche. Formerly one person had to hold a crab while a second person held the needle. The rack allowed one person to bleed six crabs and pool their blood to make a day batch. The raw lysate would then be driven to St. Louis for final processing.

"You know, I'll bet Louis Pasteur had a cleaner lab than this!"

"Yup, but nobody can argue with our results."

It was true. In today's world the lysate industry would probably have been incubated in an ultramodern, stainless steel lab fairly bristling with high-speed robotics and computer-assisted DNA-sequencing equipment. But in 1972 the lysate industry had been born in a dirty garage in Woods Hole and a storm-ravaged lab on Chincoteague Island.

CHAPTER 8

"Flugate"

[1976]

"GOOD MORNING, ladies and gentlemen. Thank you for coming on such short notice. Now, if you'll take your seats, it is eleven o'clock. We'd like to convene this emergency meeting of the Centers for Disease Control. Each of you should have a copy of a report that details the death of a young recruit up in Fort Dix, New Jersey.

"Our lab has shown that the young lad apparently died from a type A influenza similar to a virus known to be found in pigs from Asia. We suspect this strain started in poultry and was transferred to pigs through their consumption of duck feces. We have given this virus the name A/Jersey/76 H1N1."

"Good God," whispered a virologist to a colleague. "That could be the same virus that caused the Spanish flu pandemic in 1918!"

"Is that a comment for the rest of us? If not, I suggest we get started. Are there any procedural questions?"

Thus commenced the secret gathering. In normal years the purpose of the Flu Meeting is to determine which strain of flu is expected to descend on North America the following winter. Many doctors feel it is the most important annual meeting sponsored by the Centers for Disease Control (CDC) in Atlanta. But 1976 was not a normal year, and A/Jersey/76 H1N1 was not a normal flu.

* * *

Flus tend to start in Asia, then spread quickly around the world. They usually arise when farmers pick up the viruses from their poultry. Each year flus tend to be slightly different. As birds and people become ill, their flu viruses reproduce, incorporating chance mutations from two major strains of avian flu.

Scientists determine the difference between strains of viruses by looking at their outer coatings under an electron microscope. The coatings, or antigens, are made up of different types of two proteins, hemagglutinin and neuraminidase. The continuing change, or mutation, in these proteins is called antigenic drift. Hemagglutinin helps the viruses invade cells, and neuraminidase helps them burst out of cells once they have reproduced. The mixture of different types of these two proteins determines how virulent the virus will be, and the proteins are designated as H and N when the virus is given its official name. When a virologist sees the suffix H1N1, he knows he is dealing with a nasty beast.

Antigenic drift leads to a deadly game of cat and mouse. It allows the flu to stay one step ahead of the human population that has already built up an immunity to a former strain of the disease. Thus, the first order of business at the annual Flu Meeting is to determine the exact mix of the H and N proteins so that pharmaceutical companies can start producing vaccines for the new strain. This allows humans to stay one step ahead of the viruses.

Occasionally, however, something happens that is far more insidious. Instead of being passed from birds to humans, the flu virus ends up in pigs, which can incorporate RNA from both bird and human viruses to create a new, far more dangerous illness. The pigs turn an avian-human disease into an avian-human-pig disease. Doctors call this "viral sex," or shift, rather than drift. It is also the reason the CDC started to call the new strain by a more ominous name, the swine flu. The results of pig-mediated flus are dramatic. They can change the illness from a minor inconvenience to a major-league killer. They have even been known to alter history.

In 1918 the Spanish flu, a new strain of virus of avian origin, swept around the world. It killed more Americans in a single year than were killed in World War I, World War II, Korea, and the Vietnam War—and more people worldwide than died during the Black Death, smallpox, and cholera plagues. In the end, up to 40 million people died. But then the world's deadliest disease was quickly forgotten, as people tried to blot out all memory of the horrors of the world's first major modern war. Yet this flu continued to change history. Woodrow Wilson had contracted the disease and almost died. His chief aide, Colonel House, had become so ill he was unable to help Wilson lobby for enactment of the League of Nations. Some historians argue that the League of Nations could have helped avert World War II.

So it was with justifiable concern that the CDC announced in 1976 that a swine flu was poised to sweep around the world. But 1976 was also an election year. Gerald Ford was up for reelection, although he had never actually been elected. He held the distinction of being the only president ever appointed after both the vice president and president had been forced to resign from office. As an interim president, Ford had been unable to make many changes except pardoning Richard Nixon. Donald Rumsfeld led a group of advisers who feared that Ford would lose the White House unless he could deliver on a national program that would affect every citizen in a direct way. What better way to show that he could do something than to initiate a program to vaccinate "every man, woman, and child" against the evil-sounding swine flu?

For their part, the government health bureaucracies wanted to show they could play politics with the big boys. As one CDC adviser wrote, "The successful practice of public health requires salesmanship of a high order." Unfortunately, what makes for good politics does not always make for good medicine.

The federal government urged pharmaceutical firms to start churning out massive quantities of vaccine. This meant they had to inoculate over twenty million fertilized chicken eggs with swine flu. Each company had to purchase 200,000 fertilized eggs at a time and inoculate their amniotic fluids with viruses. As each embryonic chick inhaled, it drew viruses into its lungs, where they reproduced. As the embryo exhaled, the newly produced viruses would circulate back into the amniotic fluid. After two days the fluid would be so cloudy with viruses that technicians could lop off the tops of the eggs and siphon off the viruses, which would then be killed with formaldehyde and used in the vaccines.

It would take about 200,000 fertilized eggs to produce 250 gallons of vaccine using this method. Secretary of Agriculture Earl Butz is said to have reported to the president, "The roosters of America are ready to do their duty." The hens received no credit.

The program started off with a bang. President Ford announced that he would ask Congress for $135 million to inoculate every man, woman, and child before the flu season started in November. When private companies refused to insure the vaccines, Ford convinced Congress to accept liability. Remember, this was an election year; everyone wanted to jump onto the bandwagon. The media published daily pictures of thousands of people lining up to get their shots. Gerald Ford rolled up his sleeve on national television along with such notables as Henry Kissinger, Elton John, Rudolf Nureyev, and Ralph Nader. Ads were made to look reminiscent of the wildly popular polio vaccination programs of the 1950s.

But then something went terribly wrong. Ten days after the program's inception, people started to get sick from the shots. They developed fever, paralysis, and neurological disorders. Clearly, some batches of vaccine had become contaminated by pyrogens. A month after the election the program was abruptly halted. Fifty-two people had died, six hundred had been impaired, and the government faced $1.7 billion in lawsuits.

And what became of the pandemic? Nothing. Not a single case of swine flu was ever reported outside Fort Dix. It was modern medicine's most flagrant miscalculation.

There was a silver lining to this otherwise dismal tale. The four companies that had been making the vaccines were required to test the remaining stock for pyrogens. The standard test was still the rabbit test. The companies had 400,000 young rabbits on hand to run the tests. After each lot was tested the rabbits would be sacrificed, and more would have to be purchased. It was an expensive procedure. By this time, however, the Associates of Cape Cod, Mallinckrodt, the FDA, and a few smaller firms were making enough lysate so that it too could be used to test the flu vaccines for pyrogens. Initially the pharmaceutical companies were slow to use the new lysate test. But at the FDA's urging they overcame their skepticism and started to run large-scale comparisons. The results were conclusive. The horseshoe crab test was faster, cheaper, easier to use, and several times more sensitive than the rabbit test.

The following year, 1977, the FDA granted Mallinckodt and the Associates of Cape Cod the first two licenses for the commercial production of lysate and started the process of having the test incorporated into the U.S. pharmacopoeia as the standard test for pyrogens. The drug companies were now free to start dismantling their rabbit colonies. With a few strokes of a pen the FDA had created a new industry and made a single horseshoe crab as valuable as a laptop computer, a Chincoteague pony, or a bag of the finest cocaine.

And what became of President Ford? He went on to a career with the Pro-Am celebrity golf circuit. And what became of Donald Rumsfeld? He went on to earn the distinction of being the nation's youngest secretary of defense under President Ford and its oldest secretary of defense under George Bush, Jr.

CHAPTER 9

Confessions of a Horseshoe Crab Farmer

[Cape Cod, 1982]

HORSESHOE CRAB FARMING brings out an appreciation of the simpler things in life. When it is five o'clock in the morning, rain is drizzling down the back of your neck, your truck is stuck in the sand, your boat is stuck under a pier, and the tide is rising, threatening both—the first glimmer of the rising sun can give a false sense of hope. Perhaps you can just squeak out of this mess. Moments later reality returns, and with it the inescapable truth: you are up to your armpits in cold water and way over your head in horseshoe crabs.

I spent most of the summer of 1982 trying to extricate myself from such predicaments. My main preoccupation was trying to prevent my horseshoe crabs from getting wet. That's right, getting wet. Freshwater reverses a horseshoe crab's osmotic system, causing it to swell up and die. Whenever I woke to the sound of raindrops, I knew I was in for a long night of moving horseshoe crabs out of their shallow pens into deeper water—and thanks to an uncooperative volcano belching tons of sulfurous fumes somewhere over Mexico, we were having our coldest, wettest summer since sometime close to the last ice age.

I had not intended my summer to turn out this way. I had envisioned spending long sun-drenched days drifting over pellucid waters, leisurely scooping up horseshoe crabs to support my writing habit. Many writers have had similar notions—write in the morning, and pursue some simple-minded pastoral occupation in the afternoon. Thoreau had his peas, Sandburg his goats; why couldn't Sargent have his horseshoe crabs?

Indeed, why not? Several factors had led me down this particular path. First, I was broke. My teaching job at Boston College wasn't making ends meet, and the documentary film company where I had been working had just lost a major contract to produce a science series for network television. Second, I was facing the summer penniless, unemployed, and stuck in the city. Third, I had the inkling of an idea that I might be able to start a small experimental business. It would be a simple form of aquaculture, really more akin to ranching than to farming.

The project would involve rounding up horseshoe crabs when they returned from their offshore wintering grounds, bleeding them at the Associates of Cape Cod, and then holding them in pens to monitor their recovery. If all went well, the crabs could be bled one more time before being returned to the wild for the winter.

Changes in the lysate industry had created the potential for this new kind of ocean ranching. Competition was fierce. Associates of Cape Cod and Mallinckrodt equally shared about 80 percent of the market, and a handful of smaller companies vied for the remaining 20 percent. Baxter Pharmaceuticals was the exception. It was the first major firm that decided to make lysate for testing its own products but not for resale. The Illinois company had entered the business by offering to sponsor a graduate student to work in Jacob Fine's lab at Woods Hole. The move was not only good for science but good for business. Baxter was said to have saved millions of dollars in patent fees and start-up costs through this simple philanthropic gesture. Other companies had been slower about dismantling their rabbit colonies. To the old-guard animal people, the colonies had become a security blanket. They just felt it was safer to pass drug samples through a live animal than through a test tube. Plus none of the companies were sure what the FDA would do. Their fears were dispelled in 1978, when the FDA permitted *Limulus* lysate to replace the rabbit fever test.

These changes had increased the demand for horseshoe crabs. The Associates of Cape Cod now needed to bleed 250 crabs a day, or twenty thousand crabs a summer. Most of their crabs still came from Pleasant Bay, so that by 1982 a quarter of Pleasant Bay's female crabs were being bled at least once a summer. The Associates of Cape Cod claimed that no crabs died from bleeding, but I knew from our earlier work that that was not the case. We had routinely seen dead crabs being returned from the company's laboratory and knew that the Associates were having trouble finding enough female crabs to supply their needs.

So it seemed like a good idea to experiment with farming horseshoe crabs. I could make a little money, take pressure off the Pleasant Bay horseshoe crab population, experiment with aquaculture, and learn more about the lysate business. Who knew? It might even lead to a book.

I called Tom Gedaminsky, one of my students from Boston College, and we set about finding a place to hold several hundred horseshoe crabs. Eventually we discovered a saltwater pond that had been cut out of the marsh. Our neighbors had used it for several generations to hold shellfish. They gave us permission to build a fence to hold in the crabs. Eventually we had to build four more fences. As soon as the tide turned, hundreds of crabs would dig out of the sediments and surge toward the fences trying to push through, dig under, swim over or around them. More students had to be stationed between the fences to throw the escapees back into their enclosure. We christened our little operation "Limulus Farm."

But then I decided to complicate things. I knew the curator of the Baltimore Aquarium, who used to catch horseshoe crabs in Delaware Bay. When I mentioned our plan, he suggested that we buy crabs from him. This made a certain amount of sense. Horseshoe crabs range from Maine to the Yucatán, but they are most numerous and largest in Delaware Bay.

I visited the Associates of Cape Cod to see if they would be interested in crabs from Delaware Bay. Dr. Watson's eyes fairly bulged with excitement. He couldn't wait to get his hands on crabs from Delaware Bay. It meant he could save thousands of dollars, because Delaware crabs were so large he would have to bleed only half as many crabs. But there was another reason Dr. Watson was delighted. From its inception, the Associates of Cape Cod had been hampered because it was located on the northernmost edge of the range of horseshoe crabs. That had not been a problem when the company was getting started and its demand for horseshoe crabs was modest. But by 1982 the Associates needed more horseshoe crabs than Massachusetts waters could provide.

So the deal was made. John Dinga would collect the crabs in Delaware Bay and truck them to the Associates of Cape Cod, where they would be bled. Tom and I would pick them up after bleeding and drive them to Pleasant Bay, where we would hold them, feed them, and monitor their recovery.

I'll admit a had another reason for doing the project. It would allow me to see how well horseshoe crabs used for bleeding really fared under industrial conditions. It was a matter of some dispute. The Food and Drug Administration had sponsored some research in Florida, where five thousand crabs had been bled on the beach where they were captured, then released back into the water right after bleeding. The study showed that less than 5 percent of the crabs died from bleeding. But that research had been done in the wild. What would happen under industrial conditions? What would happen when you had to truck and handle them? What would happen if the bleeding line got backed up and the crabs were kept out of water all night? The summer would give us a chance to find some answers.

The project started smoothly enough. We experimented with feeding the crabs different kinds of food and discovered they loved eating scraps from the local fish market. It turns out horseshoe crabs prefer striped bass to bluefish; obviously they have a discriminating palate. We also devised a system of marking the horseshoe crabs by clipping their spines. Each clipped spine told us which week a crab had been bled. But the system had another advantage—subtlety. It allowed us to monitor crabs without the bleeders being aware of what we were doing.

But then things started falling apart. I received a late night call from John Dinga, who was driving north from Delaware Bay.

"Bill, sorry to wake you but we've got a problem. My tire just shredded on I-95. The crabs are OK, but it's going to take me awhile to get the tire repaired. Goddamn Ryder trucks—I don't think I can make it all the way out to Pleasant Bay."

"This is a bad dream right? I'm going to go back to sleep, wake up in the morning, and everything will be fine."

"Hey, at least *you* can go back to sleep!"

The next morning I woke up with what seemed like a brilliant idea. The Woods Hole Oceanographic Institution had several large empty aquaculture pools. Perhaps we could hold the crabs from this ill-fated truckload there until we could get them bled and returned to our pens on Pleasant Bay. To my relief they agreed.

Actually things had gone from bad to worse. Now we had to unload crabs from truck to pool, recapture them in the pool, load them back into our small van, drive them to the Associates, then truck them to Pleasant Bay after bleeding. And we had to be fast. I wanted to remove them as quickly as possible because there was a rumor flying around that the admiral in charge of the Office of Naval Research was planning to make an unannounced site visit to the Woods Hole Oceanographic Institution. I was not sure the admiral would appreciate seeing my horseshoe crabs in his pool.

The Associates of Cape Cod were also getting sick of the situation. Now, instead of bleeding 250 crabs a day, they had to double up their shifts in order to bleed 500 crabs a day.

Worst of all, the crabs were not doing so well either. Not only were hundreds of crabs coming back to us dead, but some of them still had needles sticking out of their carapaces. I narrowly missed stepping on a needle that was hidden under the sand where we kept the crabs. So I started to keep detailed records of the number of crabs that died from handling, shipping, and bleeding. Those that were caught in and returned to Pleasant Bay averaged 10 percent mortality. Those that were caught in Delaware Bay averaged 30 percent mortality, but on several days as many as half the five hundred crabs came back dead.

Finally I got the word from Dr. Watson. He didn't want any more of my crabs. I didn't want to give him any.

There I was: defeated, unemployed, and broke again. I decided I had to tell the town shellfish department about the debacle. It was not a pretty sight. The press had a field day. I was roundly and quite deservedly criticized as an environmentalist gone bad. Now, I am not an unbiased observer of this embarrassing situation, but I think the media missed an opportunity. Instead of using the incident as a chance to educate the public about the long-range problems of supplying horseshoe crabs to the lysate industry, they found it easier and a lot more fun to find and dis a scapegoat.

Of course the Associates of Cape Cod took their lumps as well. The FDA required that all crabs bled for lysate be returned to their native waters. But after the Delaware crabs had been bled, the Associates reportedly dumped them into Buzzards Bay, an estuary on the other side of Cape Cod that had also been depleted of its native crabs.

There was also a long-term effect. The Associates of Cape Cod became wary of working with regulators or speaking to the press. Instead of admitting to crab mortality, they denied it. Instead of explaining why they would need more horseshoe crabs, they became secretive. This would came back to haunt them. By 1985 they needed seventy thousand crabs a summer and started to buy them from Rhode Island. But this time they kept quiet about it. Their competitors would take the opposite approach.

CHAPTER 10

A Garden Party

[Charleston, South Carolina, 1985]

"CHARLESTUN IS WHERE the Coopah Rivuh and the Ashley Rivuh come togethuh to form the Atlantic Ocean," says Jim Cooper in his most elongated southern drawl. "Of course I'm no relation to the fine gentlemun. He's buried in Westminster Abbey." It's a bit of a put-on, which he performs to this day, but you can tell Dr. Cooper loves Charleston just the same.

It started when Jim Cooper was looking for a college on the East Coast where he could teach professionals how to make and use *Limulus* lysate. The inner-city campus of the Charleston College of Pharmacology proved to be the ideal location. Clouds of azalea grace its sidewalks, and great swags of wisteria tumble down its walls. An occasional palm tree reminds visitors they are in the south. The sweet smell of magnolia reminds them they are in South Carolina.

The campus was hot and humid by the time Dr. Cooper presented his course in July 1985. The lysate manufacturers had sent him samples of their best products so the class could learn the benefits and drawbacks of each preparation. The two top contenders were usually Pyrotell, made by the Associates of Cape Cod, and Pyrogent, made by BioWhittaker. This class seemed to think Pyrotell was more stable but Pyrogent easier to use. Dr. Cooper reminded the class of the summer on Chincoteague when one of his technicians removed a protective hood and a year's worth of lysate was contaminated by mold. There was no doubt about it. Making lysate was a tricky business, as much of an art as a science.

The best part of the course occurred on the last night, when Dr. Cooper invited the class to his home on James Island. The front lawn rolled down to the marsh that reflected the lights of Charleston and its harbor. Live oaks covered with Spanish moss silhouetted the garden. Guests dined on Mrs. Cooper's low-country gourmet casseroles and the succulent blue crabs that Dr. Cooper had pulled from his traps only hours before. Perhaps it was the heady aroma of magnolia, or perhaps the bellowing of nearby mating alligators, but the conversation soon turned to gossip—the main reason most people came to these meetings in the first place. Now academics, drugmakers, and FDA inspectors could let down their hair to share information on their rapidly changing industry. There was a lot to talk about.

Mallinckrodt had moved its bleeding facility from Chincoteague, Virginia, to Beaufort, North Carolina, in the late 1970s. Only a few of the original technical people had followed, and Jim Cooper was not among them. The division ended up being run by the wife of a shrimp fisherman.

In the meantime the parent company had become embroiled in one of those merger and acquisition frenzies that seem to sweep through American business on a regular basis. Management, who had let Mallinckrodt's financial position deteriorate, sold its pharmaceutical divisions to the cosmetic giant Revlon. Revlon, however, quickly decided it should concentrate on its core business and forced Mallinckrodt to sell its patents, bleeding facility, and lysate inventory to Whittaker Bioproducts. But BioWhittaker, as it came to be called, soon found out that the patents were insufficient, the facility run-down, and the inventory unstable. Biowhittaker had to recall two years' worth of Pyrogent, and lawsuits and counterlawsuits followed.

These machinations had put the Associates of Cape Cod in the driver's seat. They called BioWhittaker's customers to convince them to change brands and quickly captured 80 percent of the market. But they still had a less than user-friendly product.

By late evening the conversation at the Coopers' garden party turned serious. "So Jim, why don't you get into the business? You know how to make a better product. You've been in the industry from the beginning. Why not give it a try?"

"I dunno," mused Dr. Cooper. "I've thought about it, but it seems like a big jump to go from the comfortable environs of academia to the brutal jungle of business. They might eat me alive."

"But look at Stan. He just stumbled into the business blindly, and look how well he's done."

"You could make a killing."

"Or I could lose my shirt."

But the idea niggled in the back of Jim's mind for several months. Dr. Cooper knew that BioWhittaker faced several basic problems. It was located in Walkersville, Maryland, a hundred miles from the nearest horseshoe crab and a nine-hour drive from the run-down bleeding facility it had acquired in Beaufort, North Carolina.

Bill MacCormick could remember that drive well. He had just been hired by BioWhittaker, and his job was to drive the crude lysate from North Carolina to Maryland. "I had to tighten the springs on my car so they could hold up under the weight of three chests of ice and twenty gallons of lysate. Somebody should have been riding shotgun. The future of BioWhittaker was riding in the back of that car, along with all my worldly possessions."

So having separate bleeding and processing facilities had put BioWhittaker at a disadvantage, because it made it twice as expensive for them to produce lysate as for their main competitor, the Associates of Cape Cod, who bled their crabs and produced their final product in the same plant. But separating the dirty business of bleeding from the sterile process of manufacturing final lysate had a hidden advantage: it cut down on contamination. In the long run, this would prove to be a crucial advantage. And there was another benefit to having separate facilities. It was better for BioWhittaker's basic raw material—horseshoe crabs. They could be bled on-site and put back in the water immediately after bleeding. This cut down on the number of crabs that died from the bleeding.

If Jim started a lysate company in Charleston, he felt he would have several advantages over both of his main competitors. First, he would have to bleed only half as many crabs as the Associates of Cape Cod, because South Carolina crabs were bigger and held twice as much blood as crabs from New England. Second, he could separate the bleeding and manufacturing operations in two buildings at the same location, unlike BioWhittaker. Third, and most important, he knew how to buffer his lysate to make it more stable than BioWhittaker's Pyrogent and more user-friendly than Pyrotell, made by the Associates of Cape Cod.

By 1986, Jim had finally decided that the time was ripe. McGaw, a major producer of blood products, announced it was going to close a plant in Milledgeville, Georgia. Jim drove down to the auction and was able to fill a tractor-trailer truck with centrifuges, autoclaves, ventilation hoods, and glassware. "It allowed me to start on a shoestring. You could never do it today. The FDA wouldn't allow it. One of its inspectors once told me, 'Pharmaceuticals is no place for the little guy.'" But in 1986 an entrepreneur could still slip through the starting gate. Jim incorporated Endosafe in September and received financial backing in 1987.

But then disaster struck. One of the hazards of the lysate business is that you want to be located near horseshoe crabs. In the United States, that means the East Coast. But the East Coast is also home to the world's most damaging storms—hurricanes. Jim had discovered this once on Chincoteague, but he would have to discover it again in Charleston.

Most hurricanes circle their opponent, landing only glancing blows. But Hugo in 1989 was different. It was born off Africa, grew strong in the West Indies, and took direct aim at Charleston. "Our house took a licking, and Endosafe was damaged. But what really hurt was that the port of Charleston was closed. We were shut off from our supply of horseshoe crabs for several days. It made us realize how vulnerable we were. Endosafe almost went out of business. Of course the Associates called our clients again and offered to sell them their lysate. I kidded them a few years later when they were hit by Hurricane Bob, asking the Associates if they wanted me to call their clients to offer them our lysate."

After the 1989 storm, the Coopers made two adjustments. Jim called New Jersey to see if he could line up a backup supply of crude lysate, and Frances started petitioning the South Carolina legislature to have horseshoe crab fishermen be licensed. She started by educating the public. State officials had discovered that 20 percent of the horseshoe crabs caught in South Carolina waters were being shipped to Virginia to be used as bait by conch fishermen. Frances also had to find sponsors for the bill and line up people to testify. But on the day the bill was supposed to be debated, Jim had to be out of town.

"Jim, what am I going to do?"

"Well, why don't you speak to them? You know as much about horseshoe crabs as anyone."

"But, I'm not a scientist."

"Neither are they. You'll do just fine."

She did. The legislators were enchanted with Mrs. Cooper's testimony. They had never heard of this lucrative new fishery, where you could actually put the fish back in the water after you were done with them. The legislature passed the bill, and the governor signed it just in time for the 1991 season. The regulation stated that licenses would be given only to fishermen who were not going to adversely affect the population of horseshoe crabs. In reality, that meant that horseshoe crabs could be used only for lysate.

The law had teeth as well. After it went into effect, Jim caught one of his suppliers in the parking lot selling horseshoe crabs to an out-of-state fisherman. The crabs had already been bled and were supposed to be returned to South Carolina waters. But the out-of-state fisherman had other ideas. He was going to truck them to Virginia and sell them as bait to conch fishermen. In other states the supplier could have been paid handsomely for his double-dipping. But in South Carolina he lost his license.

PART III

Environmental Conflicts

CHAPTER II

Fishing for Bait: The Conch and Eel Fisheries

[Delaware Bay, 1990–2000]

HUMANS HAVE USED horseshoe crabs for almost as long as the two species have coexisted. Native Americans fertilized their fields with horseshoe crabs and fashioned spear points from the crabs' tails. Colonial farmers followed the Native Americans' example, using horseshoe crabs for fertilizer and chicken feed.

It wasn't until the mid-1800s that the use of horseshoe crabs for fertilizer and chicken feed became a bona fide industry. As part of his postdoctoral work, Carl Shuster chronicled the Delaware Bay industry by collecting archival photographs of the practice. The photos showed that fishermen on the Delaware side of the bay could simply pluck the mating crabs off the beach and stack them like cordwood. Each neat pile contained hundreds of thousands of rotting corpses. On the New Jersey side of the bay fishermen had to catch horseshoe crabs in offshore pound nets. Then they tossed the crabs willy-nilly into seventy-foot-long wooden holding pens that held eighteen thousand crabs at a time. On both sides of the bay, the piles of dead and dying horseshoe crabs raised such a stench that neighbors complained, and chicken eggs smelled like old fish.

Place names are believed to reflect the old practices: Murderkill River, Dead Horse Creek, King Crab Landing, and Slaughter

Beach. If they do not refer to the old industry, one shudders to contemplate the alternative explanations.

More than four million crabs were killed in 1870, but then the fishery started to decline decade by decade until it totally collapsed in 1950. The industry was stopped in the 1960s not because of what you might hope for—concern for the animals—but because of the smell of rotting horseshoe crabs. (Environmentalists have learned to be philosophical about such things: if odors prove stronger than science, argue smell over numbers.) After the fertilizer companies shut their doors, the population of horseshoe crabs rebounded so well that by 1977, when David Attenborough arrived to film his award-winning television series *Life on Earth*, there were more than enough horseshoe crabs mating on Delaware Bay to be suitably impressive.

But by the 1980s horseshoe crabs were facing a new threat. During the spring and fall, blue-crab fishermen and oyster dredgers had traditionally set out eel pots to make a little off-season money. Their favorite bait was large egg-laden female horseshoe crabs, so the fishermen would fan out on the beaches of Delaware Bay during the new- and full-moon tides of April and May to collect the spawning crabs. It was an extremely easy task. Hundreds of thousands of crabs would simply crawl up to the fishermen's boots in their frenzy to lay and fertilize their eggs. The fishermen could collect more than enough crabs to fill their needs. Half of the crabs would be chopped into pieces and stuffed into eel pots to catch "shoestring" eels in the spring. The rest of the crabs would be frozen to be used in the fall to catch silver eels as they prepared to migrate out of Delaware Bay on their way to their spawning grounds in the depths of the Sargasso Sea.

Initially the eel fishermen had concentrated primarily on this fall migration. The eels would be shipped to Italy and other Mediterranean countries where eels are a traditional Christmas delicacy. But eels became a lucrative spring fishery when Asian sea farmers started to import live American "shoestring" eels to supplement and stock their aquaculture pens.

What fishermen called "shoestrings" are actually six-inch-long jet-black elvers that have just metamorphosed from their leptocephalus larval stage. The transparent, leaflike larvae look so unlike adult eels that for several decades scientists thought they were an adult species of plankton found only in the Sargasso Sea. The leptocephali drift in the grip of the Gulf Stream until they are swept close to the American coast. Here they develop rapidly into elvers and follow the odor of freshwater back into places where they originally hatched such as Delaware Bay. If all goes well, they swim up the bay and spend long lives in freshwater and brackish water tributaries. If things go poorly, they make an unfortunate detour into an eel pot and end up being flown to Asia for a considerably shorter life in an overcrowded aquaculture pen.

Bobby Bateman was one of the first fishermen to take advantage of the horseshoe crabs' usefulness as bait. He is an energetic New Jersey oysterman who is popular with his crew and shares their healthy disrespect for authority. His seventy-five-foot boat is rigged for dredging oysters. The heavy chain-link dredge bites several inches into the ocean floor and scours up whatever lives there. Three of his crew members are kept busy culling oysters from the debris. But that is not all they catch. Sometimes the dredge returns to the surface bristling with the tails and flailing legs of horseshoe crabs.

"By the end of a good trip the boat will be piled so high with horseshoe crabs we look like a haystack steaming into port. On land, I have seven or eight more guys that load the crabs into tractor-trailer trucks and drive 'em back to my freezers. For a long time I was the only guy in the business, so I sold horseshoe crabs to eelers from Maine to Florida and to most of the catfishermen in the Chesapeake."

But in 1993 another New Jersey fisherman entered the market. He was Charles Burke, who operated an offshore fishing trawler. His boat was rigged to skim a trawl over the ocean floor to catch bottom fish like cod, haddock, flounder, fluke, and whiting. But by the 1990s most of the fisheries for these fish had collapsed, so federal regulators were encouraging fishermen to target underutilized species like horseshoe crabs, conch, and eels. Charles needed no such incentives. He started catching horseshoe crabs in his trawls and marketing them to conch fishermen.

"Conch meat became the yuppie food of the nineties. All of a sudden there was a worldwide market for conch meat sushi. You could truck live conch to restaurants all over this country and air-freight conch meat to markets in Japan and China. It sure as hell beat the old scungilli market for the local Italian restaurants."

For a few years Charles and Bobby had almost all of the mid-Atlantic horseshoe crab fisheries to themselves. But Charles had an advantage over Bobby. His boat was registered to fish in federal waters, and he could land in any major port to off-load his horseshoe crabs. This would become useful when environmentalists started pushing for restrictions.

In the early 1990s New Jersey officials started hearing reports of northern conch fishermen coming from as far away as New Hampshire. They would arrive in rental trucks, collect horseshoe crabs on the northern side of Delaware Bay, and speed back up I-95 in darkness. During the mid-1990s the conch fishery expanded, and Delaware officials started to see southern fisherman traveling north from Virginia and Maryland to collect horseshoe crabs on the southern side of Delaware Bay.

In 1996 environmentalists convinced New Jersey governor Christie Todd Whitman to cut the catch of horseshoe crabs and ban daytime collecting to thwart out-of-state fishermen. This started an elaborate shell game. Charles Burke began catching horseshoe crabs in New Jersey waters and steaming south to Maryland to unload them. There his crew would pack them into trucks and drive them back to New Jersey for resale. Maryland attempted to stop the practice by limiting the catch to 1,500 crabs per boat, so Charles started off-loading his crabs in Virginia and driving them back to New Jersey to sell. As Virginia came into compliance, Charles started off-loading first in Philadelphia, then New York. At $30,000 to $40,000 a boatload it was still worth it to pop up in a different port every trip to foil the bureaucrats. Besides, it was pretty good fun to boot.

When federal regulators finally created a 1,500-square-mile horseshoe crab sanctuary off Delaware Bay, Charles was quick to claim some pride of authorship. "They created that sanctuary just to stop one guy—me. It was Christie Whitman's pet project from the beginning. I think it was payback time. A bunch of us fishermen staged a protest at one of her political rallies when she was running for reelection. We were there with pickup trucks and placards yelling our heads off. It certainly ruined her whole day. As they say, payback is a bitch. Well, the bitch paid us back!

"There's a safety issue too. I can't even steam through that sanctuary with a load of crabs. I have to steam thirty miles to get outside one side of the sanctuary or fifty miles to get outside the other. Those are treacherous waters out there. Even Delaware has cut me off from transporting crabs through the C and D Canal. They've regulated me out of the fishery. That's what they've done. It's draconian I tell you, draconian."

In a sense Charles is right. The sanctuary was established to curtail the behavior of Charles Burke and Bobby Bateman. But as flamboyant as the two were, they were not alone. By 2000 over 2.3 million horseshoe crabs were being harvested every year for bait, and the East Coast population was starting to decline, as it had in the 1800s.

There is a certain irony in all of this. A horseshoe crab is worth exponentially more over its lifetime when used for lysate than if it was killed and chopped up for bait at 75¢ a pound. Some states consider bait to be a good use for horseshoe crabs, because they still regard horseshoe crabs as shellfish predators. But this seems to be economically beside the point. Cooked shellfish is worth about $5 a quart, but a quart of processed horseshoe crab is worth over $15,000.

Prices fluctuate as demand changes. People used to catch lobsters so they could chop off their tails to chum for striped bass. I don't think I know anyone who is still dumb enough to continue that outdated practice. Do we want to do the same with horseshoe crabs?

Of course the government is partly to blame for this situation. It is a classic case of the left hand not knowing what the right hand is doing. The FDA should have spent the last twenty-five years educating the public about the economic value of horseshoe crabs to human health. Instead the Department of Commerce was giving market advice and low-interest loans to encourage fishermen to catch "underutilized" species like horseshoe crabs, conch, and eels. Environmentalists have learned that one of the best way to determine what animal will soon be in danger is to check what animal is presently being touted as an "underutilized" species.

CHAPTER 12

A Day at the Beach: Red Knots and Horseshoe Crabs

[Reed's Beach, New Jersey, May 29, 1985]

BRIAN HARRINGTON has studied shorebirds so long that he has started to look like a red knot. He has a twinkle in his eye, a wry sense of humor, and long, bushy insectivorous eyebrows. He explains his early interest in shorebirds laconically: "It all started with a BB gun on the shores of Rhode Island." Later, he supplemented his intuitive feel for bird behavior with courses at Ohio Wesleylan and the University of Southern Florida.

But it was in 1972, at the Manomet Bird Observatory in Massachusetts, where Brian found his life work. For a long time ornithologists had been puzzled about the whereabouts of certain birds during migration. They knew where the birds spent their winters and summers but were mystified how they got between the two points. It seemed impossible that such small animals could store enough energy to fly directly from South America to the Arctic. Brian had a hunch that migrating shorebirds landed at certain discrete beaches where food was so plentiful that the birds could bulk up quickly before undertaking the next leg of their flight. He remembered the beaches of his childhood and wondered what would happen to the shorebirds if these critical areas were developed.

Shortly after starting his first job, Brian had a chance to try out his idea on Archie Hager, a trustee of the observatory and the Massachusetts state ornithologist.

"I think you're just plain wrong. But if you are correct, the best bird to work on would be the red knot. Every fall we see thousands of them on Cape Cod Bay, but then they just disappear. We suspect they fly to South America, but we can't account for over a million birds."

"Wow!"

"You could probably write letters to every birder on the East Coast and see if any of them has ever seen large concentrations of red knots."

Brian did send out three hundred letters but received only a single, lukewarm response.

"Dear Mr. Harrington, we've seen a few thousand red knots here on Reed's Beach, but I don't think it would be worth your while to come down. They're not very interesting birds."

The following spring Brian and his assistant, Linda Leddy, met their discouraging correspondent at the post office in Reed's Beach, New Jersey. It was cold and gray as they piled into their informant's open pickup truck and drove five miles down the long hardscrabble beach. Intermittent patches of sand and mud flashed by the window as their driver warned them not to have high expectations. Finally they drove through a line of ramshackle summer cottages nestled into the low dunes.

"We couldn't believe our eyes," remembers Linda. "The flats were alive with birds. You couldn't even see the beach. It was shoulder to shoulder with bobbing, rippling, feeding shorebirds. Most were chestnut-colored red knots, but occasionally you'd also see the crossed scissor wings of laughing gulls poking into the air. The plaintive whistles of the shorebirds were matched by the raucous calls of the gulls. There must have been a hundred thousand birds and a million more beyond our line of vision. We were totally overcome, but the driver couldn't understand our excitement. These were just common shorebirds, not exotic species that you could add to your life list. He had no idea this was unusual. He thought the entire East Coast probably looked like this in early spring."

Brian returned to Reed's Beach every spring and started banding the birds wherever he could find them. Gradually he pieced together the puzzle. Red knots spend the summer above the Arctic Circle and the winter on the windswept beaches and hardpan lowlands of Tierra del Fuego. In March they hopscotch up the Argentinean coast to Brazil, where they pause to feed on snails and *cochina* shells in the shallow *pannes* of Lagoa do Peixe. They double their weight in a month, then fly eight days directly over the Atlantic Ocean to the beaches of Delaware Bay. They land thin and emaciated, but this will not last for long. They have timed their migration to coincide perfectly with the phases of the moon and the breeding of horseshoe crabs.

Almost all the red knots in the world spend the next two weeks on these beaches. Each bird eats 135,000 horseshoe crab eggs; together they gorge down 248 tons of fat and protein. Each square meter of the beach contains over a million eggs, totaling more than a pound. But the birds can reach the eggs only in the top six inches of sand. They must rely on there being so many horseshoe crabs that late-arriving crabs dig up the eggs laid by earlier-arriving crabs. And there still must be another million eggs buried eight inches below the surface to ensure the continuation of the crabs. It is superabundance on a staggering scale.

The horseshoe crab eggs give the red knots the fuel they need to fly directly to the Arctic and start laying eggs. The horseshoe crab eggs are critical to the birds' survival. When they arrive above the Arctic Circle, there is no food. The ground is still frozen and covered with snow. They may not eat for two more weeks.

Brian's work has been adventurous but not easy. He has banded red knots in Massachusetts, Argentina, and Brazil. He banded birds sitting side by side in Massachusetts and found them seven months later in Argentina still sitting side by side with a flock of a hundred thousand birds. He has found the same birds coming back to exactly the same beaches to feed on precisely the same patches of plentiful food.

One year Brian and Linda banded two hundred knots in Argentina, flew back to Boston, drove down to New Jersey, and saw the same birds waiting for them on Reed's Beach. "I was still exhausted from just sitting on the plane," remembers Linda. "Imagine how tired the birds were."

Brian and Linda recollect that their most memorable day at Reed's Beach occurred in 1985. May 29, to be exact. "The tide was dropping, and we were heading into dusk. Most of the crabs had crawled back into the water, but the odor of roe still hung in the air. The beach was olive green with billions upon billions of tiny green horseshoe crab eggs. The beach was more eggs than sand." But the birds were not eating or flying. Usually at this time of day they would be flying inland to spend the night. But the knots seemed more alert and vigilant than usual. Their posture was somehow different. Brian knew something was about to happen. Then suddenly it did.

At 5:15 sharp a thousand birds took off at once. Their wings sounded like the wind of a passing gale. They kept rising higher and higher until they disappeared, heading north-northeast. Two minutes later a thousand more birds followed. "It was like watching old footage of Vietnam, where wave upon wave of planes were launched every few minutes to carpet bomb the Ho Chi Minh Trail. In a few hours it was mostly over. We had witnessed the exodus of over fifty thousand birds."

"The following day the beach was empty. All that remained were dead crabs and a few thousand plover. It was just amazing to think of those red knots way up there, winging their way toward the Arctic.

"We took a compass bearing of the birds as they flew north-northeast. Fourteen hours later we received a call from Chris Rimmer on Hudson's Bay. He had just seen red knots flying overhead on exactly the same course. Those were our birds. I know it in my bones."

CHAPTER 13

The Atlantic States Marine Fisheries Commission

[Washington, D.C., 1996–1999]

NEWS FLIES QUICKLY through the birding world. Brian Harrington started publishing his findings in journals and popular magazines. Television producers picked up the information and broadcast stunning footage of horseshoe crabs in David Attenborough's BBC series *Life on Earth* and a *NOVA* film, "The Sea behind the Dunes." *Life on Earth* featured nighttime footage of horseshoe crabs spawning on Delaware Bay, and "The Sea behind the Dunes" showed horseshoe crabs mating on the clean white sands of Pleasant Bay. Both productions captured the awe of witnessing this ancient ritual.

I wrote a book while working on the *NOVA* film and was astounded at the number of letters I received from people who wanted to see horseshoe crabs on Pleasant Bay. It was no wonder. The confluence of mating horseshoe crabs and the migration of a million shorebirds is truly one of the wonders of the natural world. Many people find it as impressive as the mating of salmon in Alaska or the migration of wildebeest in Africa.

Witnessing horseshoe crabs mating on a silent beach in spring is like visiting your own private Paleozoic Park—a far older and more majestic experience than Jurassic Park, that parvenu of a park hatched out of the bloody brain of the fertile writer Michael

Crichton. It is like visiting our planet before there were birds, mammals, trees, or flowers, to say nothing of those devilishly clever dinosaurs. Yet this ritual occurs within a hundred miles of every major East Coast city. All you have to do is get to the right beach at the right time. A good naturalist can help. Audubon societies were quick to capitalize on horseshoe crabs' newfound fame by sponsoring expeditions to Delaware Bay and the Monomoy Wildlife Refuge on Cape Cod.

For several summers I supported my writing habit by leading horseshoe crab cruises on Pleasant Bay. Besides providing income, much needed in those early years, they allowed me a way to keep an eye on what was happening to the crabs. What became obvious was that we were not alone on these cruises. Passengers were dismayed to see collectors dumping crabs into floating carboys and a sixty-foot pen hidden in a marsh behind the islands. On Delaware Bay, birders watched as fishermen waded into flocks of shorebirds to collect horseshoe crabs while they were still mating. It was becoming apparent that collecting mating horseshoe crabs was going to have an impact on both the crabs and the shorebirds that depended on them.

In 1986 Brian Harrington proposed that a series of reserves be established to protect the most crucial areas where shorebirds stopped to refuel. Eventually the system of reserves would stretch from Argentina to the Arctic, but the first site to be selected was Delaware Bay. The reserve didn't provide for any new regulations, but what it lacked in legal weight it made up for with moral heft. Soon so many birders were visiting Delaware Bay that more than $10 million a year was clinking into the registers of cash-strapped local communities.

But fishermen were strapped for cash as well. In Delaware fishermen resented efforts to keep collectors off the beaches during daylight, when red knots were feeding. In New Jersey fishermen resented efforts to keep collectors off the beaches at night, when out-of-state fishermen could work undetected. Everyone was upset that there were no reliable figures on the actual number of crabs in Delaware Bay.

Into this vacuum stepped Jim Finn, a man with a unique perspective. As a marine biologist and co-owner of a small company that produced lysate, he realized that collecting horseshoe crabs for both lysate and bait could have a serious impact on the future of Delaware Bay. In 1990 he started coordinating a group of volunteers to survey spawning horseshoe crabs during the full- and new-moon high tides in May and June. During the next decade, over 150 volunteers would spend several thousand man-hours counting horseshoe crabs. The surveys provided an annual snapshot of Delaware Bay's overall population of spawning adult crabs.

The results were instructive. During the first two years the surveyors estimated there were more than 1.2 million crabs. This was certainly down from the years when that many crabs were being harvested for fertilizer, but it showed a decent population probably able to both support both red knots and produce the next year's crop of new horseshoe crabs. But then the numbers started to plummet: 400,000 in 1992 and only 100,000 in 1995. And that was not all. The numbers of eggs in the sand and the numbers of offshore crabs were declining as well. However, one number was still rising: the number of crabs being caught for bait. By 1996 New Jersey fishermen harvested 900,000 horseshoe crabs, almost a third of Delaware Bay's total population.

It was clear that a crisis was imminent. When the spawning season arrived in 1997, the scientists were ready. In early May they reported that fewer red knots were on their regular beaches in New Jersey. Apparently, fewer birds had arrived from South America, and they found so few eggs in New Jersey that they had flown to the Delaware side of the bay, where collecting had already been restricted. By mid-May scientists were becoming more confident that their numbers were real. There really were fewer crabs and fewer birds on both sides of the bay. The crisis was at hand.

Scientists and environmentalists banded together to urge New

Jersey governor Christine Todd Whitman to take action. The pressure paid off. On May 30, halfway through the spawning season, Governor Whitman declared a thirty-day moratorium on harvesting horseshoe crabs. Environmentalists were elated, fishermen furious. It made for good political theater. Newspapers urged her to declare a second moratorium after the first expired. Finally Governor Whitman made the ban permanent by making it illegal for New Jersey fishermen to dredge or trawl for horseshoe crabs in Delaware Bay.

The ban should have put Charles Burke and Bobby Bateman out of business, but it didn't. They presented their case to the New Jersey Marine Fisheries Council, which overturned Governor Whitman's regulations. Conservationists banded together again and threatened legal action. The council backed down, and the ban held. But federal waters were still unregulated. Charles Burke started to catch crabs in federal waters, off-load them out-of-state, and drive them back to New Jersey for resale. For $20,000 a trip the extra driving made sense. The practice was not strictly illegal, but it had the appearance of circumventing the intent of the law.

It was during this time that another environmentalist became involved. Gerald Winegrad was a young up-and-coming legislator from Maryland. "I was on a train going to a meeting about Delaware Bay when I happened to read a fact sheet about red knots and horseshoe crabs. I decided then and there that I had to go over there to see for myself." Winegrad was hooked. In 1996 he took a job with the American Bird Conservancy in Washington, D.C., and convinced them to join a coalition of environmental organizations petitioning to have the entire East Coast closed to the fishing of horseshoe crabs.

There was no way the petition would succeed, but it did get the attention of the Atlantic States Marine Fisheries Commission. The ASMFC is a strange beast. It was conceived in 1942 to help states on the East Coast manage their fisheries. It came into prominence when it successfully brought about the recovery of striped bass up

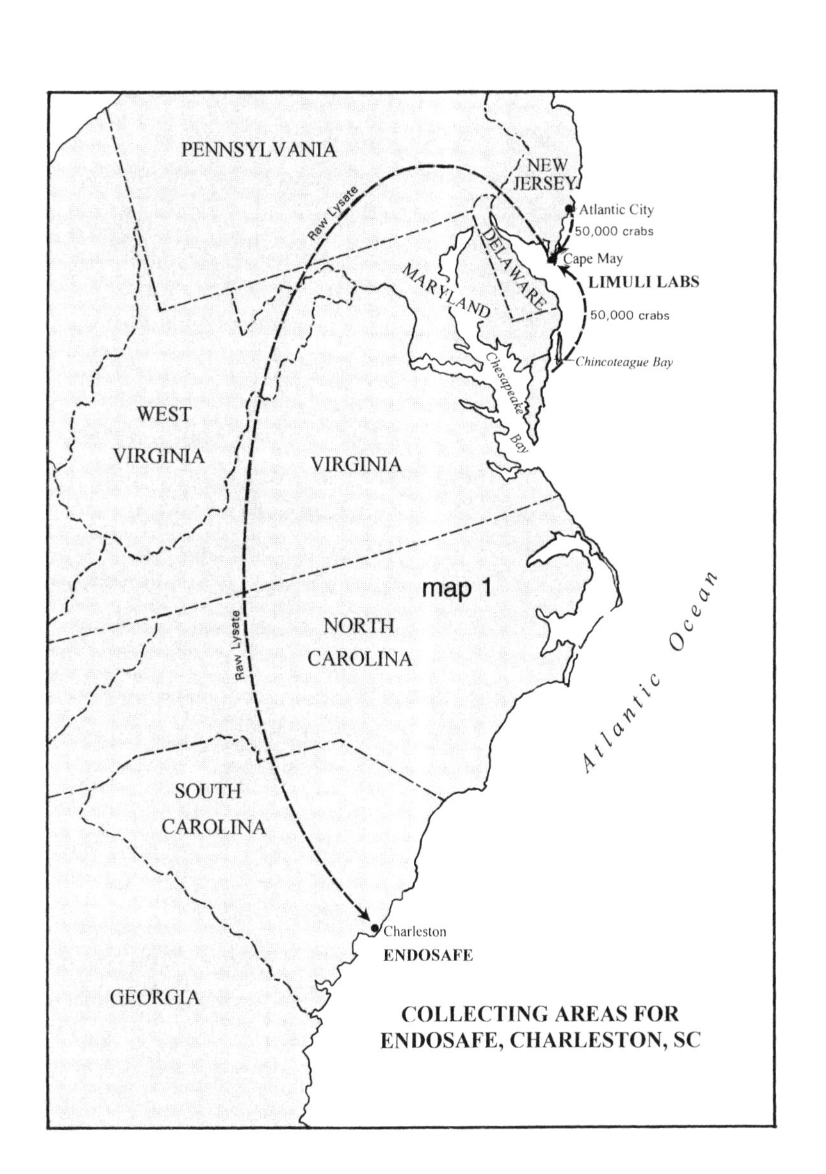
PENNSYLVANIA
NEW
JERSEY
Atlantic City
50,000 crabs
Raw Lysate
Cape May
LIMULI LABS
DELAWARE
MARYLAND
50,000 crabs
Chincoteague Bay
Chesapeake
Bay
WEST
VIRGINIA
VIRGINIA
map 1
NORTH
CAROLINA
Raw Lysate
Atlantic Ocean
SOUTH
CAROLINA
Charleston
ENDOSAFE
GEORGIA
COLLECTING AREAS FOR
ENDOSAFE, CHARLESTON, SC

and down the East Coast. This has been hailed as one of the most successful environmental efforts of the past fifty years.

But striped bass had an ardent following of recreational fishermen willing to make sacrifices for the sake of their favorite fish. Horseshoe crabs had only a few weird friends and legions of enemies ready to accuse them of everything from eating clams to being ugly. Plus the environmentalists were up against a group of commercial fishermen who wanted to continue making easy money for little work.

Gerald's first meeting with the Atlantic States Marine Fisheries Commission did not go well. The people who ran the commission were not impressed with the National Audubon Society or the American Bird Conservancy. They didn't have the time or money to listen to a bunch of birders. The commission was established to protect the livelihoods of fishermen, not to heed the gripes of shorebird fanciers.

Never underestimate the persistence of birders. They will compete like hell to identify the first spring warbler, but when it comes to an environmental threat, watch out. They have members, organizations, and e-mail alerts. The commission did not realize any of this. Within weeks they were deluged with petitions, e-mails, even letters from Congress. The commission relented; perhaps they could fold the horseshoe crab problem into a management plan for eels. But that was not good enough. More lobbying ensued, and in 1997 the commission voted to take on the issue of horseshoe crabs.

The first thing the commission had to do was find out how many horseshoe crabs each state was actually catching. The commissioners agreed that each state could either take an average of its catch from 1995 to 1997 or report its catch for 1999.

The results were astounding. Some states were unaware that they even had a horseshoe crab fishery; others were ignorant of its size. Rhode Island officials reported a harvest of 26,000 crabs. They were unaware that more than 60,000 crabs were shipped into Massachusetts every year to be bled for lysate, then shipped back to Rhode Island to be sold for bait. Massachusetts officials had their own problems. First they had to get a handle on how many horseshoe crabs were actually being caught. How did they do this? They asked fishermen. Of course the permit holders knew that the numbers were going to be used to set quotas, so they lied like hell. Anyone who had ever held a license doubled, tripled, even quadrupled the number of crabs caught—then added in Uncle Louie, who used to catch a few crabs so he could have a little scungilli for Christmas. When asked to list "location of capture," license holders provided such useful information as "in the water" or "none of your Goddamn business." It was all pretty good fun, tweaking bureaucrats' tails. Massachusetts ended up reporting 400,000 crabs, which was about 300,000 too many.

But, for what they were worth, the commission had its numbers. The next order of business was to figure out what to do with them. That would be decided at the commission's fall meeting to be held in Mystic, Connecticut, on November 3, 1999.

Michael Dawson drove down from the Associates of Cape Cod to argue that the lysate industry should be exempt from any regulations, because they killed only 10 percent of their crabs. It was the standard party line. Gerald Winegrad reported that he had seen a New Jersey boat unloading 10,000 crabs in Virginia. A gentleman from Massachusetts reported that he had seen fishermen with Virginia license plates filling up trucks with horseshoe crabs. It was clear that a lot of crabs were being shuttled around. Environmentalists praised New Jersey, Delaware, and Maryland for getting the ball rolling. Those states had cut their catches by as much as 60 percent, and the conservationists argued that the other states should do likewise.

By the end of the meeting the commissioners agreed that each state would reduce its catch by 25 percent. They also recommended that the National Marine Fisheries Service prohibit the catching of horseshoe crabs in Charles Burke's favorite spot, the federal waters off Delaware Bay. After the meeting, the commissioners had to return to their home states to decide how to implement the reduction.

The April 2000 meeting proved even more contentious. Virginia announced that it could not comply with the commission's management plan and had established a cap of 710,000 crabs, up from the 200,000 crabs the state's fishermen had landed in 1999 and half a million crabs over their quota. This was a shocker. Virginia was announcing it was still open for business. It would continue to let out-of-state fishermen off-load crabs caught in other states or in federal waters. As a result, all the other states' efforts had been in vain. Their fishermen could simply sell those states' crabs to Virginia's $7 million conch industry. Virginia still intended to be the "Loophole State."

Gerald Winegrad made the first comment from the public. "Mr. Chairman, I think the commission should reject Virginia's management plan and hold them out of compliance."

After an hour of debate the chairman called the vote to declare Virginia out of compliance. If the declaration passed, the secretary of commerce could shut down Virginia's horseshoe crab fishery for as long as five years.

"Maine."

"Yes."

"New Hampshire."

"Yes."

It continued that way down the fifteen East Coast states to Florida. The final tally was fourteen yeas, one nay. Virginia was notified it would be out of compliance by the first of May.

Before the meeting closed, the National Marine Fisheries Service reported on its plans to declare a 1,500-square-mile area off Delaware Bay closed to horseshoe crab fishing. This would close the other loophole, but without Virginia's compliance it would also be ineffective. Nerves were getting frayed. When a member of the public proposed that the closed area be called the Carl Shuster

Horseshoe Crab Sanctuary, a representative from the National Marine Fisheries Service responded. "I'm glad to hear the suggestion made by Mr. Oates, because if this is named the Carl Shuster sanctuary, then I'll know where to send the letters of complaint after we implement it."

The meeting closed with a call to write a letter to the FDA "to see if we can respectfully get a copy of their permits and licenses specifically regarding the collecting and distribution of horseshoe crabs post bleeding."

The meeting adjourned at noon on April 4, 2000. The battle to protect horseshoe crabs had just begun.

CHAPTER 14

Jay Harrington vs. Bruce Babbitt

[Pleasant Bay, 2000]

THE SUN IS SETTING on Pleasant Bay. A tall bearded man hunches over the stern of his boat, releasing bled horseshoe crabs back into the clean shallow waters. Jay Harrington chose this simple, almost zenlike life twenty-five years ago.

After finishing college, Jay moved to Cape Cod to seek a simple life close to nature. For several years he lived on a small island in Pleasant Bay, where he and his two brothers potted for eels, listened to jazz records, and tended a small patch of marijuana plants. Early one morning the Orleans Police Department staged an amphibious raid on the island. Officers roused and shackled the sleeping brothers and seized their marijuana crop. Photos of the raid were splashed all over the local newspapers. There could hardly have been a better example of governmental overkill than the arrest of the mild-mannered brothers who just wanted to be left alone. It was treated as a local joke but was humiliating to those concerned. After the arrest the Harringtons became hardworking model citizens.

In 1976 Jay started collecting horseshoe crabs for the Associates of Cape Cod. It was good, physical, outdoor work, the kind of summer job a student might have to help pay for college tuition. But the work provided Jay with a good living. He was able to make $100,000 for four months' work. Then he would take the winter off to volunteer as a disc jockey at a radio station in Provincetown. Over the years he built up a substantial audience who liked his laid-back style and extensive collection of jazz records. Jay also became known as a defender of horseshoe crabs. He made a point of taking local reporters and environmentalists out in his boat to explain his operation and show how carefully he treated his crabs.

Jay became so good on camera that the Associates of Cape Cod realized he could help them project a softer, greener image. They started sending newspaper and television reporters to interview Jay. Soon he was making appearances on the Discovery Channel and Japanese television. Jay found he enjoyed the exposure and in later years would often list his media credits to establish his credibility. He became so adept that he was referred to as Dr. Harrington on the docket papers that announced his trial.

Things started to change for Jay in the mid-1980s. The Associates wanted more crabs than Pleasant Bay could supply. Jay decided to expand his operations to Monomoy Island. But Monomoy was different from Pleasant Bay. Fishermen had learned to recognize Pleasant Bay as "Jay's territory" and avoided catching "Jay's" horseshoe crabs. This was the kind of self policing that managers encourage. It helps avoid "the tragedy of the commons." But Monomoy was a different story. Other fishermen already used the area to catch crabs for bait and lysate.

In 1991 Jay met with Bud Oliveira, the regional administrator of the Monomoy National Wildlife Refuge. Jay used the meeting to complain about the number of crabs the bait fishermen were catching. Evidently he was persuasive. Bud gave Jay the only permit to collect crabs in the Refuge and banned bait fishermen outright. Not only did Jay get the permit, but the permit stipulated that he was required to turn in any fishermen he saw poaching for horseshoe crabs. Bud saw it as a way to buy an extra pair of eyes on the water. Jay saw it as a way of protecting his supply of horseshoe crabs. Bait fishermen saw it as a way of giving an unfair monopoly to an out-of-town competitor.

As the mid-Atlantic states started tightening their regulations, the town of Chatham felt the repercussions. Suddenly there was a new market for local horseshoe crabs. Fishermen could make good money catching crabs off Monomoy and trucking them south for resale. During the summer of 1999 people in Chatham started to see boats loaded to the gunwales with horseshoe crabs. At night fishermen would land the boats and hurl thousands of crabs into tractor-trailer trucks. It didn't help that the boats were called Carolina skiffs and the rental trucks had license plates from Virginia. Rumors flew through the town. People resented that their local horseshoe crabs were being sold out of state.

Bob Barlow had seen all this before. He was an MBL neuorophysiologist who had spent his career studying vision in horseshoe crabs on a beach near Woods Hole. However, in 1993, he had been forced to suspend his research because collectors had removed too many crabs from his field site on Buzzards Bay. Some nights he had to pay the fishermen $500 or $600 so he could return the crabs to the water. It came out of his National Science Foundation grant earmarked for his world-class research. "Your tax dollars at work," Bob chuckled at a public forum in Chatham. But the payments were not enough. Dr. Barlow had been forced to move his lab to a former lighthouse in Chatham in order to be near to enough horseshoe crabs to continue his work.

In 1999 Bob was ready to resume his studies. He had invited a German documentary film crew to go out with him on the Monomoy flats. While the crew was waiting, they heard that there were three boats moored nearby filled with dying horseshoe crabs. Bob's landlord, Deborah Ecker, confronted the fishermen, who claimed they were going to drive the crabs to Rhode Island to be bled for lysate. The documentary crew captured everything on film, including the lie about bleeding the crabs for lysate. There was no lysate facility in Rhode Island.

Mrs. Ecker described the incident at the board of selectmen's meeting, established the "Save Our Horseshoe Crabs Committee," and started writing letters. One of the fishermen lost his license, and Bud Oliveira, the Monomoy Refuge administrator, decided to take another look at his policies on horseshoe crabs.

That was one of the purposes of the Atlantic States Marine Fisheries Commission: to help managers like Bud know what was going on in other parts of the country so they could decide what to do in their own backyard. Bud started doing some research. Bud learned that the Chincoteague National Wildlife Refuge had turned down requests for permits from lysate collectors wanting to catch horseshoe crabs within its borders. He contacted a lawyer from the Department of the Interior who determined that Jay's permit should never have been issued, because it was out of compliance with the organic act that established the Refuge. Evidently "organic act" meant something quite different to a lawyer than to someone more biologically inclined. Bud decided to rescind Jay's permit because it had never existed in the first place. It was all a little embarrassing.

Bud's counterpart at the Cape Cod National Seashore was equally chagrined. Maria Burks was the newly appointed superintendent of the Seashore. "I was totally embarrassed," she admitted. Her staff had been following the stories in the local press and now also read the ASMFC reports. Gradually it dawned on them that Jay had been removing horseshoe crabs from the Seashore for twenty-five years. "We couldn't believe our eyes. We spent several months checking to see if a former superintendent had given him some kind of permission. What were we going to do, start chasing Jay with our fleet of pursuit canoes?"

Maria had a problem. The eastern side of Pleasant Bay lies within the Cape Cod National Seashore, and it is illegal for anyone to make a profit from wildlife in a national seashore. Furthermore it is illegal for any person "to possess, destroy, injure, deface, remove, dig or destroy fish or wildlife, their remains or eggs in a national seashore." Jay was in violation of at least four of these provisions. He had been illegally collecting crabs in National Seashore waters for over twenty-five years.

Both Bud and Maria separately announced that they were going to ban the practice of catching horseshoe crabs on the Monomoy National Wildlife Refuge and in the Cape Cod National Seashore. Jay felt he was being picked upon unfairly. "I'd always been the one who said these animals needed to be protected."

It was like the marijuana raid all over again. But this time Jay had friends in high places. Senator Ted Kennedy invited state and federal officials into his private office to see if they couldn't work out a compromise. Jay turned to the Associates of Cape Cod.

A lot had changed at the company. It was no longer the small entrepreneurial firm that Stanley Watson had started twenty-five years before. Stan had no heirs, so he had decided to sell the firm to a company that would keep it in the forefront of the lysate industry. There had been considerable interest when the Associates went on the market. Both competitors and former customers made bids. The company was courted by six major US pharmaceutical companies for several years. Stan was said to have thoroughly enjoyed the attention, but in the end he opted to sell his firm to Seikagaku, a major Japanese pharmaceutical firm that got its start harvesting shark cartilage for an alternative medication for arthritis. The sale went through in 1997 and was finalized on January 1, 1999. It gave the Associates access to plenty of money in case of emergencies. Loss of access to horseshoe crabs was just such an emergency.

In 1999 Tom Novitsky, the new president of the Associates of Cape Cod, now a wholly owned subsidiary of Seikagaku-Japan, placed a call to Peabody and Arnold, the old-name Boston law firm Stan had always used to threaten his adversaries. Jay Harrington was about to find himself suing Bud Oliveira, Maria Burks, and Secretary of the Interior Bruce Babbitt; in addition to the Department of the Interior, the U.S. Fish and Wildlife Service, the

National Park Service, the Cape Cod National Seashore, and the Monomoy National Wildlife Refuge.

The last time Jay went up against such odds, he lost. What would happen this time? The Associates of Cape Cod filed their case on March 22, 2000.

CHAPTER 15

A Bizarre Incident

[May 23, 2000]

THE ASSOCIATES OF CAPE COD went on the offensive after filing their case. Time was short. Judge Rya Zobel would likely make some sort of ruling before the horseshoe crabs returned to spawn in May.

The first thing that the Associates wanted to do was create a positive public image. They invited reporters to visit their production facility and urged environmental organizations to hold public forums. ACC's message was twofold. First, they were the good guys, and the bait collectors were the bad guys. Second, southern fishermen were sneaking into Massachusetts to steal the commonwealth's horseshoe crabs.

But something about their message didn't ring true. I hadn't thought much about horseshoe crabs for several years, but from what I remembered of the situation in Pleasant Bay they were wrong on both counts. Their collectors were the only people catching crabs in the bay, and despite the rumors, nobody was coming into the bay from out of state. So I decided to write an article that I hoped would help put the issue in perspective.

But nothing added up. The Associates were saying that they were bleeding only 60,000 crabs; that didn't make sense. It would mean that they were bleeding fewer crabs than they had in the mid-1980s, yet they had continued to expand and had added seventy more employees to bleed crabs. Where were those crabs? Then

I remembered that in 1985 the Associates had requested permission to collect 35,000 crabs from Rhode Island. How many crabs were they collecting from there now? Nobody would tell me.

Then I got lucky. I had called an employee of the company about tagging horseshoe crabs, and he happened to mention that during the peak of the season the Associates bled 1,700 crabs a day. I had to hold my voice steady to finish the conversation. Now it was simply a matter of math. They had to be collecting at least 60,000 *additional* crabs from Rhode Island in order to have enough to bleed 1,700 a day.

The Associates had failed to mention to the public that almost half of their crabs came from Rhode Island. Officials in Massachusetts and Rhode Island seemed to be equally unaware of the situation. In fact Rhode Island found out about the extra crabs only at an Atlantic States Marine Fisheries Commission meeting, when Massachusetts delegates told them.

However, the main thrust of my article was that the federal government should not close Pleasant Bay and Monomoy, because this would put pressure on smaller areas less able to support collecting for lysate. I had concluded by saying that the population of horseshoe crabs in Pleasant Bay seemed to be healthy after twenty-five years of collecting.

Not everyone agreed with that observation. In fact, while writing the article, I engaged in several arguments with my old friend George Buckley. George insisted that the population of immature horseshoe crabs in Pleasant Bay had collapsed. This was based on the numbers of cast-off shells he had been counting for the past twenty years. During the late summer the immature crabs would shed their shells, which collected in the dried eelgrass of the wrack line, where George's students would count them. "But George, there are plenty of crabs in Pleasant Bay. I've seen 'em. How could the immature crabs be declining? You must have been going out at different times every year, or perhaps your techniques just stink!"

Still, something about my first article's conclusion bothered me. The numbers people were reporting just didn't make sense. Take Pleasant Bay. If the Associates had been bleeding 40,000 crabs a summer and they admitted to a 10 percent mortality, that would be 36,000 crabs left after the second year, 33,400 after the third, 30,060 the fourth, 27,000 the fifth, . . . 16,000 the tenth, 5,559 the twentieth, and only 3,283 adult females left after twenty-five years of collecting—and George was claiming there was no replacement by immature crabs! By anyone's numbers, there should be virtually no adult female crabs left in Pleasant Bay. But undoubtedly there were. In fact the bay was chock full of horseshoe crabs.

Then I received a call from a friend who reminded me that the summer before I had pointed out the unusual number of large crabs in Pleasant Bay. I had even joked that someone must have been introducing alien crabs. Suddenly it hit me. That was it! Somebody really had been introducing crabs into Pleasant Bay. George and I were both right. The number of immature crabs had been declining, and yet the population of adult crabs had been increasing, because the adult crabs had been introduced, probably for several years!

But first I had to prove to myself that the crabs really were larger than I remembered from previous years. This would be difficult, because neither George nor I had measured the crabs in any systematic way. Then I remembered reading a paper by Carl Shuster. He had measured crabs up and down the East Coast. Perhaps he had figures from Pleasant Bay. Carl told me the size of the crabs over the phone and graciously promised to send me his field notes.

I bought a meter stick and drove down to Pleasant Bay. It was March, so the crabs hadn't arrived, but there were still shells on the beach from the summer before. I started to measure the shells. I couldn't believe my eyes. Out of the first twenty I measured, six, over a fourth of the crabs, were too large to come from Pleasant Bay. I was trembling with the excitement that a scientist must feel when he realizes that he knows something nobody has ever known before. Granted, I had not discovered string theory or a new nuclear particle—but I trembled just the same.

It would have been a perfect discovery if I had found that something in nature had caused such a change. Unfortunately it indicated that someone I liked had been introducing the crabs into Pleasant Bay. This was a tricky thing for me to handle. I realized that Jay Harrington had probably started introducing crabs for the same reason I had, to replenish the stocks in Pleasant Bay. But I had done it only for a month; it looked like Jay had done it for as many as ten years. If this were true, it meant that not only had horseshoe crabs been illegally removed from Pleasant Bay but they had been illegally introduced as well.

Of course, no one was going to believe such a disturbing scenario based on results from only twenty crabs. I called Troy Hopkins, an old friend who taught environmental science at a nearby high school. His class would be able to measure five hundred crabs, the same number that Carl had measured in 1951. Then I called the *Boston Globe* to see if they wanted an article for their science section.

It was not my job, but in writing this second article I leaned over backward to be fair to the Associates of Cape Cod. I interviewed both Carl Shuster and John Valois from the MBL. John said: "I don't think there is a soul who could could go down to Pleasant Bay and point to evidence that the estuary has been damaged in any major way. Biologically there may be nothing wrong with transplanting those crabs. We don't see any species that you would expect to see increase has increased, and we don't see any species you would expect to have declined has declined." Carl Shuster said: "I think the feds should at least give the company a period of grace to line up other sources of crabs."

If that version of my article had run, the Associates would have gotten off the hook almost scot-free. Here were two of the top experts in the field saying they doubted that the ecology of Pleasant Bay had been damaged by the introduced crabs and ACC should be given a chance to find other supplies of crabs.

But then I received a strange phone call. It was from my editor at the *Boston Globe*: "Bill we have a problem. I just got a call from the publisher. Apparently ACC got to his wife, and now he wants me to kill the story. I told him I believed your version and don't want to kill the story but said that I would talk to you about having another writer interview ACC to get their side of the story. Would you agree to that?"

What should I do? I didn't have time to argue. The case might be decided any day, and I didn't have time to write an article for another publication. So I agreed. The other writer interviewed the Associates, and my editor rewrote the story, taking out all references to the introduced crabs. The article ran on May 23.

Judge Zobel granted ACC a temporary injunction the next day. Her ruling gave ACC permission to collect crabs for one more season in the Cape Cod National Seashore and on the Monomoy Wildlife Refuge. The case would be continued in the fall.

I had to teach a marine biology course that summer, so I wouldn't have time to write any more articles. But I felt it was important to get the whole story to the public so that other people could see for themselves—the crash of the local crab population, the illegal importing of larger alien crabs, and the cover-up, including the ACC's fudging of numbers. Time was short, so I committed the journalist's greatest sin—I gave my scoop to another writer!

Doreen Leggat had already been approached by the Associates of Cape Cod to visit their lab and go out with Jay Harrington. But they didn't expect her to ask any difficult questions. She wrote a special series that ran for two weeks on the cover of the *Cape Codder*. Her specials were far more damaging to Seikagaku than my original piece would have been. If my original article had run in the *Globe*, Seikagaku could have quietly admitted that perhaps some crabs had been introduced into Pleasant Bay but that the top experts in the country didn't believe any damage had been done, and it wouldn't happen again. Instead they were put in the uncomfortable position of having to deny that the crabs had ever been admitted at all.

The series received wide attention and elicited numerous letters to the editor. But the revelation that seemed to upset the Associates most was that both my article and Doreen's series had mentioned that ACC was now owned by Seikagaku.

I had written about the new ownership not because Seikagaku was a Japanese firm but because I felt it important for people to know that ACC was no longer a local enterprise and was now owned by a large multinational corporation. The Associates seemed to be embarrassed by their parent company. Their lawyers never mentioned Seikagaku in the court case. Seikagaku was not even written on the sign outside their Cape Cod plant. This seemed curious and somewhat disturbing. By this time both Endosafe and BioWhittaker had also been bought by larger firms and seemed proud of their affiliation. Endosafe was now officially Charles River Endosafe, whose parent company was Charles River Labs of Boston, and BioWhittaker had been bought by Cambrex in Chicago.

Evidently the articles were a serious blow to Seikagaku's public relations initiative. They canceled all visits and refused to talk to the press when the court case reconvened in the autumn.

CHAPTER 16

The Decision

[John Joseph Moakley Federal Court House, Boston, Massachusetts, May 22, 2001]

"GOOD MORNING, Judge."

"Good morning Lisa. Good morning Jay. Is that the memorandum of decision?"

"That's it."

"Good. I'll go over it while I'm getting into my robes."

Judge Rya Zobel took the document from her assistants, closed the door of her private office, and looked out over Boston Harbor. She loved this view. Her office was on the top floor of the new federal courthouse on Boston's Fan Pier. The imposing building had a sweeping parabola of tinted glass windows that took in the entire view. Below her a flotilla of sailboats dashed between ferries disgorging passengers from Logan Airport and the south shore suburbs. Occasionally a ferry would have to pause to avoid one of the whale-watching boats departing from the New England Aquarium. It was a warm day. Interns and paralegals sprawled on the grass busily punching numbers into cell phones and chatting about restaurants, dates, and a fabulous new cocktail dress that was a total steal at only $2,000. It was a decidedly urban setting, an incongruous venue for such a decision on such a decidedly rural issue as horseshoe crabs.

Cape Codders had been waiting for Judge Zobel's decision

for almost a year. Lawyers had requested a summary judgment in November, then February, then March. Now it was May, and Jay Harrington was already in full operation collecting crabs in the Cape Cod National Seashore and on the Monomoy National Wildlife Refuge. But Judge Zobel had discovered that this was a surprisingly complicated case. It involved two federal areas governed by three separate acts of Congress. Last spring Judge Zobel had felt she didn't have enough information to make a well-thought-out decision, so she had granted ACC a temporary injunction to continue harvesting and a year's time to line up an alternative source of horseshoe crabs.

Since the injunction, the case had been argued in a stream of lengthy documents. However, there had been no public trial, and neither side had spoken to the press. It was clear that everyone wanted to settle the case without the circus atmosphere of the year before.

The case had always been about jurisdiction. The federal agencies wanted to be able to regulate horseshoe crabs as wildlife, but the Associates of Cape Cod wanted horseshoe crabs to be regulated as shellfish by Massachusetts officials. Massachusetts had traditionally regulated shellfish in federal waters and leaned toward the rights of fishermen. So the first question that Judge Zobel had decided she must face was whether horseshoe crabs should be categorized as shellfish or wildlife.

"Shellfish" was a centuries-old category dating back to the time when anything living in the water was considered to be a fish; thus we still have anomalous names like starfish and jellyfish. "Shellfish" was an everyday subcategory of "fish" that included crabs or clams with shells: crustacea and mollusks, in scientific nomenclature. But since those early days, researchers had discovered that horseshoe crabs were neither crabs nor crustacea, let alone shellfish. They inhabited their own little subclass of spiderlike creatures called Merostomata. The reason that this distinction was so important was that when Congress established the Cape Cod National Seashore it ceded the management of shellfish to local towns but made it illegal to take wildlife for profit.

Judge Zobel smiled at her law clerk's use of language. "It does not seem that Congress had any opinion about the designation of horseshoe crabs . . . because the statutes and legislative history do not mention them." Furthermore, "There was no indication that Congress meant to protect the gathering of horseshoe crabs as 'a distinctive pattern of human activity,'" as ACC had argued. Finally, "Congress certainly would not have considered the large-scale commercial bleeding of horseshoe crabs for the manufacture of LAL [*Limulus* amoebocyte lysate], because it did not exist at the time of the seashore's inception."

So much for divining the intent of Congress, a dicey proposition under the best of circumstances. Such are the fine points that law clerks must consider, but what Judge Zobel had found most surprising was the lack of any discussion of what was really best for horseshoe crabs.

In fact what Monomoy and the Seashore had proposed was exactly what scientists had found to be the most effective method to restore marine wildlife: the creation of small reserves to protect animals during spawning. They had discovered four more things as well. One, that fishermen would fight such marine reserves with both tooth and nail. Two, that after only a few years the number of fish in the reserves would double and spill out into the adjacent waters. Three, that fishermen would quickly find these waters and line up at daybreak to reap the benefit. And four, that only the most ornery fisherman would deny that the number of fish had doubled outside the boundaries of the reserves as well as inside.

Federal officials had always wondered why the Associates of Cape Cod had sued them in the first place. They had always been willing to accommodate Jay, and the reserves had never posed a real threat to his livelihood. He could have continued to make a decent income collecting outside the boundaries of the proposed

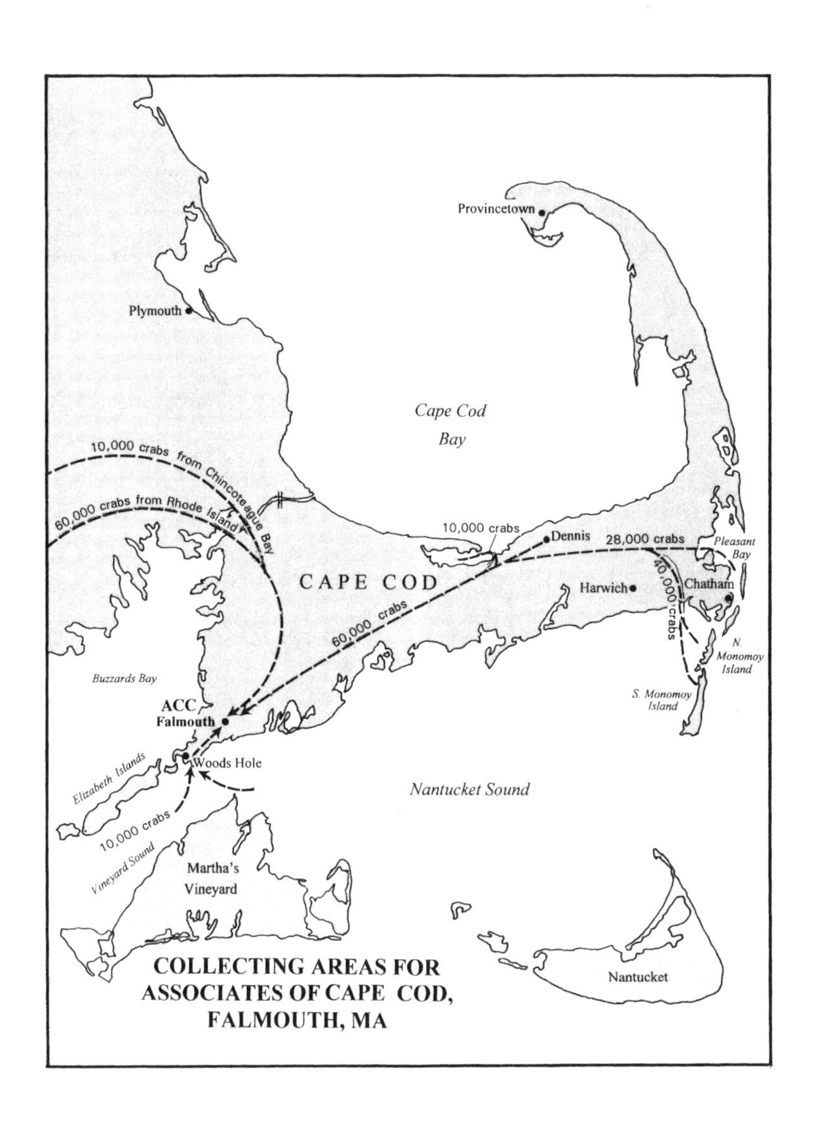

COLLECTING AREAS FOR ASSOCIATES OF CAPE COD, FALMOUTH, MA

reserves. In fact, in the long run, Jay Harrington and the Associates stood to benefit more from the reserves than anyone else. The reserves would have both increased their catch and protected their primary supply of horseshoe crabs. However, Jay and the Associates had decided to fight the reserves.

Judge Zobel had responded with a Solomon-like decision. She had instructed her clerks to cut the case down the middle, allowing the National Seashore to create a reserve in Pleasant Bay but allowing the National Wildlife Refuge to create a reserve only in the parts of Monomoy covered under the Wilderness Act. This still allowed the Refuge to create a reserve in 90 percent of their waters. Judge Zobel had made her ruling on a legal technicality. She deemed it arbitrary and capricious that the Refuge had decided that its reasons for giving Jay a permit to collect crabs had been faulty. In effect, it had changed its mind in the light of new evidence—that sort of thing is encouraged in science but frowned upon by lawyers. However, Judge Zobel had given the Refuge the option of undertaking a new study to determine whether collecting horseshoe crabs was compatible with Monomoy's mission to protect shorebirds. Finally, Judge Zobel had given both sides ten days to decide whether they wished to appeal her decision.

Judge Zobel finished reviewing the document, buttoned up her robe, and strode into court. She liked her decision. It was just, fair, and well grounded. Apparently, the Associates agreed, or they thought it would be too expensive to disagree. They never filed to appeal her decision.

CHAPTER 17

The Loophole

[Falmouth, Massachusetts, June 2001]

THERE IS LITTLE to distinguish the building that houses the Associates of Cape Cod. The Falmouth Police Station and a graveyard guard its eastern flank; restaurants and real estate firms are its immediate neighbors. The building's fake Tudor architecture looks as if it still houses the Steakery restaurant; the simple wooden sign makes the Associates of Cape Cod look like the mom-and-pop real estate firm it used to be. Perhaps this is by design.

Yet the site is not inviting. A chain-link fence overtopped with barbed wire surrounds the entire perimeter of the property. No sign anywhere mentions that the building houses the headquarters of Seikagaku-America.

There is little evidence that losing the court case has slowed Seikagaku's expansion in any way. The company is forging ahead with plans to build a $20 million facility in the Falmouth Industrial Park. It will provide greater security and room for the hundred workers that will be employed during the peak bleeding months.

The first truck arrives at the company's padlocked gate. It is Jay Harrington, delivering crabs from Monomoy. He waves at a second, unmarked, Seikagaku truck. It is heading out to pick up crabs from the Aquaculture Research Corporation. The crabs have been dredged out of Rhode Island waters and delivered to the Dennis facility. The third truck is a refrigerated truck from "A and A," a major fish-processing company in Fairhaven. It is delivering crabs trawled from Vineyard Sound and landed in Woods Hole.

The fourth truck is a tractor-trailer from Virginia. The driver is Leon Rose. He used to catch crabs for BioWhittaker, but last year the Associates made him an offer he couldn't refuse. Now he is catching crabs in Virginia and driving five hundred miles north to sell them to Seikagaku. But the arrangement isn't working very well. Leon has made several trips north, and each time the Associates have refused to pay for crabs damaged by trawling or transportation. More crabs have died on the way back to Virginia. It seemed ridiculous to throw the dead crabs back into the water.

Leon is losing patience. He bought a larger boat and outfitted his tractor-trailer with refrigeration and a system of nozzles to spray the crabs, all based on what he thought was a gentleman's agreement. He delivered ten thousand crabs during the spring months when the crabs were unavailable in New England. Now the Associates are saying they don't want his crabs, because northern crabs are again available. It has to be the money. New England fishermen can afford to sell crabs for $2 a head, but he has to cover the expenses of his boat, his truck, his employees, and the time he spends delivering the crabs and waiting a day for them to be bled. He had planned to return the crabs to Virginia, but it simply makes more sense to resell them for bait. It is no longer illegal to do so.

While the crabs come and go as usual, a lot has been going on behind the scenes at Seikagaku. During the early days the Japanese owners were content to let their American managers run the company. But the company lost the court case, decided not to appeal, and the Food and Drug Administration was still threatening to shut them down. It was clear they needed new direction and new people. Tom Novitsky's legal strategies had been ill advised, and his media campaign had blown up in the company's face. Seikagaku had been embarrassed by having the company's name mentioned so prominently in the newspapers. The Japanese managers hoped to solve these problems by making several changes.

The easiest problem to solve would be contamination. The Japanese managers felt they could overcome the company's ongoing contamination problems by continuing plans to move out of their cramped old building in downtown Falmouth. The new industrial park building would allow them to separate the bleeding and manufacturing operations and hire thirty new employees.

Their second problem was how to find enough crabs to replace the crabs they used to get from Pleasant Bay and Monomoy. The managers decided to address that problem with some inside moves. First they convinced the FDA to reverse the long-standing requirement that lysate companies had to return horseshoe crabs back to the sea. This had been easy. The FDA had always harbored a special fondness for the industry they had helped to create. Second, the Associates went to Massachusetts officials to support what became known as the "rent-a-crab" policy. The policy allowed fishermen to sell crabs for lysate, then turn around and resell them for bait. It had been proposed by a bait fisherman who wanted to enter the lucrative lysate business. The new policy was a biological disaster. Massachusetts had only a few small areas with enough crabs to support the lysate industry. The areas had done so for twenty-five years, but this policy could wipe them out in just a few short months. It would convert what had once been a sustainable live crab fishery into a dead-end, dead bait occupation. But it didn't matter to Seikagaku if crab stocks were depleted in Massachusetts as long as they could increase their catch from out of state. That realization had led to the third inside move: convincing the Atlantic States Marine Fisheries Commission to allow states to transfer their horseshoe crab quotas to other states.

Massachusetts was in a particularly good position to transfer quotas. Its quota was artificially high because it had overreported its catch for 1999. Now it had a cap of 300,000 crabs, although its fishermen had caught only 128,000 in 2000. That gave Massachusetts 172,000 crabs it could transfer to another state. Who wanted those crabs? The Virginia conch industry. This was the beginning of an unholy alliance between Massachusetts and Virginia, the rogue state that the ASMFC had voted out of compliance on May 1, 2000.

The three strategies had opened a huge loophole. The Associates could now buy crabs from mid-Atlantic states, and their fishermen could resell the crabs back in their home states or in Massachusetts. Environmentalists feared the loophole would lead to a domino effect that would deplete horseshoe crabs farther and farther south. Other lysate producers feared that soon all four companies would be fishing from the same pot.

The Associates gave up all pretense of caring for the future of horseshoe crabs after losing their case against the Interior Department. But their concern had been suspect for at least ten years. The former director of research at the Associates, Norman Wainwright, had always believed the company line that ACC cared for their crabs and treated them gently. But one day the company's regular driver had been ill, so Norman and a colleague had volunteered to drive some bled crabs to the Aquaculture Research Corporation to be returned to the water. When they arrived at the holding pens, Norman and Steve Boyd were surprised to see a bunch of fishermen's trucks lined up to meet them.

"What are you guys doing here?"

"We're waiting for you."

Steve and Norman looked at each other in disbelief. The fishermen were there to buy the crabs right off the truck. Obviously this had been going on for a long time and had been kept secret from some of the most highly placed scientists in the company.

It had not ended there. Seikagaku sued Dr. Wainwright when he started working for the MBL.

Jim Cooper, the founder of Endosafe, also realized the implications of the new policies. In June he penned an article for the *LAL Times*, an industry newsletter published by Endosafe. The article warned that the new policies could result in a cap being put on the collection of horseshoe crabs for biomedical purposes because the policies would "blur the stark difference between bait fishermen and lysate producers and allow the biomedical industry to be perceived as just another source of the horseshoe crab decline."

The June article was politically astute, but it came out too late. When quota transfers came to a vote before the Atlantic States Marine Fisheries Commission in April, three states—New Jersey, Delaware, and Maryland, the three mid-Atlantic states that had done the most to protect horseshoe crabs—joined the U.S. Fish and Wildlife Service in voting against the measure. But the other nine states voted in favor. The regulation passed with the loophole intact.

CHAPTER 18

Raw Lysate: A New Industry

[Chincoteague, Virginia, July 30, 2001]

IT IS ALMOST MIDNIGHT. The lights of Ocean City, Maryland, twinkle in the still night air. The captain has steamed several hours to reach this spot. He wipes his brow. "I'm gettin' too friggin' old for this nighttime shit," he sighs. "Can't fish during the day in this heat, too damn hot for us and too damn hot for the crabs. Sun'd dry 'em out, and BioWhittaker wouldn't pay us a nickel for them."

But now the tepid moon is rising slowly over the horizon. It is drawing the incoming tides across the shallows of Greater Gull Bank. This is the signal the horseshoe crabs have been waiting for. They will clamber out of the mud and start searching for food. They are hungry. High winds have kept them holed up in the sediments for several days.

The captain gives the word, and a seventy-foot net slides over the stern to scour the bottom. Thirty minutes later it is dangling over the deck, bulging with a gelatinous mass of eyes, fins, and writhing legs. A crew member slits the cod knot, releasing a collection of slimy creatures worthy of a Hieronymus Bosch print. They clatter to the deck and slither toward the nearby scuppers.

The crew start to cull the evening's catch. Sharks, skate, and jellyfish are shoveled back over the side, but the conchs and mud-covered horseshoe crabs remain. Hundreds of the large crabs lie on their backs, vainly thrashing their legs in the air. A few clamber

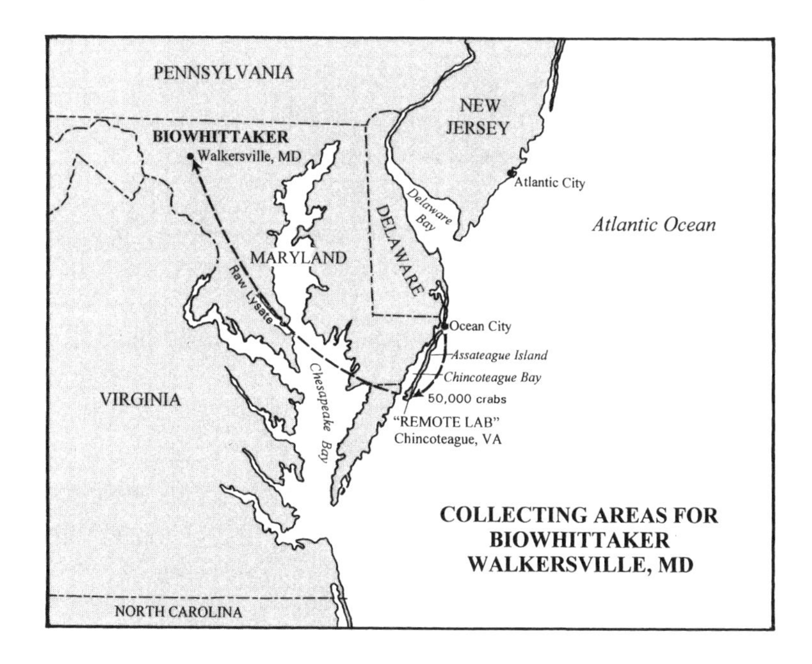

over each other in a slow-motion effort to escape. But the crew members work fast. They discard the small crabs, toss the large ones into plastic totes, and cover them with wet burlap bags. The bags will keep the crabs moist for the long trip back to Virginia.

"A few more hauls like that and we should be home in time for a couple hours of shut-eye. But remember boys, nobody gets paid until we load these suckers back into the truck and deliver 'em to Jan. A thousand crabs ought to keep her busy for the day."

By 5:30 Jan Nichols is already on her way to the "Remote Lab." Someone at headquarters dubbed it "Remote" because he thought Chincoteague seemed so distant from BioWhittaker's main facility in Walkersville, Maryland. Somehow the name just stuck. But the "Remote Lab" suits Jan just fine. She prefers to run her own show —close to the ocean and far from the corporate types.

Chincoteague is at its shining best this morning. The sun dances and glitters off the bay, and a pair of egrets stalk through the marsh grass hunting for minnows. A flock of gulls jostle raucously for bits of bait left on the causeway by last night's fishermen. Despite the heat Jan drives with the air conditioner off and one window open. She wants to be the first person on the island to hear the gabblings of snow geese as they tumble out of the sky after their short summer in the Arctic.

Jan stops. Day-trippers are starting to trundle over the causeway with boats in tow. Farther up the road the commercial guys are gathering at the Shore Stop to gossip, joke, and gulp down coffee. Jan knows better than to stop there. Most of the guys at the convenience store know her, but the island has an unwritten law that the commercial fishermen meet at the Shore Stop and the recreational fishermen meet at Barnacle Bill's. Sure the guys would be polite, but it would break the early morning code, the illusion that Chincoteague still belongs to the local working stiffs, not outsiders looking for Chincoteague ponies.

Jan has seen more than her share of Chincoteague ponies. Her lab lies on the route the ponies take to the auction block. During pony week fifty thousand extra tourists crowd the narrow streets. For the last few years she has just closed the lab and given everybody the week off to avoid the hassle of pony week.

Down one of these roads is the chicken house where Jim Cooper and Ed Hockstein started the lysate industry thirty years ago. Jan has always marveled at how she stumbled into the business. She grew up in Michigan and spent most of her life as an engineer. But like many midwesterners she had always harbored the romantic ambition to live on the ocean and study marine biology. The divorce allowed her to live her dream. She gave herself three choices: live on either Key West, Cape May, or Chincoteague. Her move to Chincoteague had made all the difference.

In 1988 the local paper ran an ad for someone to run a marine lab on Bunting Drive. Jan had never heard of horseshoe crabs, *Limulus* lysate, or BioWhittaker, but they hired her anyway.

Since then Jan has grown to love horseshoe crabs. "I look at these unlikely animals every day, and it's like going back in time. No other creature has contributed so much to science and to human health without being killed in the process. To me they are just perfect animals, unchanged for hundreds of millions of years." But it wasn't until doctors discovered her intestinal tumor that Jan truly learned how perfect horseshoe crabs really are. As she lay in bed fighting the cancer, Jan felt the presence of a powerful ally. "I knew that every drug that dripped into my arm had been tested with horseshoe crab blood. It made me feel protected and somehow reassured. Today there is not a day that goes by that I'm not thankful to be working beside such an amazing creature—the animal that helped save my life from cancer."

Jan swings into BioWhittaker's ranch-style building in a residential section of Chincoteague. It is 6 A.M.

"Good morning, Tom. Did you enjoy your week off?"

"Sure did. Thank God the tourists have gone. Did the ponies leave any presents on their way to the auction?"

"Manure? I hope not. By the way, who's here? The crabs should be arriving anytime. It's been so hot the guys have been catching 'em at night. As soon as the crabs arrive, have some of the temps wash the mud off their shells and get them ready for bleeding. Can they all do the bleeding now?"

"Yup, they're coming along just fine. By this time last year all of our college students would be drifting back to school."

"Yeah, I was talking to Benji last week. She's tearing her hair out. Her summer help are all leaving, and she's expecting her fifth child in September."

Benji Lynn Swan ran Limuli Labs up in Cape May, New Jersey. Although they kept trade secrets to themselves, Jan and Benji were still pretty good friends. Both of them had carved out an interesting niche in the *Limulus* lysate industry, providing raw lysate to larger firms that would then freeze-dry the final product and package it for resale. Separating the two processes was better for the crabs and cut down on contamination. It also provided an interesting business opportunity. Benji owned her own company and sold raw lysate to both Endosafe and Haemechem in St. Louis. Jan was a paid employee of BioWhittaker's now owned by Cambrex in Chicago.

At seven o'clock eight hundred crabs arrive and are washed and prepared for bleeding. Soon cobalt-blue blood, as precious as gold and as healing as modern medicine, is frothing and foaming into dozens of empty beakers.

Jan knows that every year her collectors have to work a little harder and travel a little farther to catch the one hundred thousand crabs that BioWhittaker needs to stay in business. This summer alone, her collectors had gone several weeks without catching any crabs at all. In public, the lysate industry blames bait fishermen for depleting the crabs. But Jan admits that the biomedical industry has to shoulder some of the blame as well. In the early years her fishermen collected crabs by hand on the beaches of Assateague. Later they fished from trawlers, but they still caught the crabs on their way in to the beaches to spawn.

Jan also knows that instead of returning the Maryland crabs to Maryland waters her collectors would simply dump them into Chincoteague Inlet. It made a certain amount of sense. Some of the crabs would crawl back into Tom's Cove to replenish the stocks, so the following season the Virginia collectors wouldn't have to pay for fuel to steam back to Maryland to recapture the same crabs. But in fact most of the crabs released in the inlet would be recaptured for conch and eel bait. It was an inside thing. The collectors and the bait fishermen all laughed about it at the Shore Stop. The practice has been mostly stopped, but not before hundreds of thousands of crabs were removed from the spawning populations off Maryland and Delaware Bay.

In many ways the horseshoe crabs are like the Chincoteague ponies. The ponies used to sell for $20 a head and attract a few hundred outsiders. Now the ponies sell for $2,000 a head and attract fifty thousand extra tourists. The lysate industry has grown the same way. Collectors used to hand-gather a few hundred crabs from local beaches; now they have to roam far and wide to catch almost half a million crabs for a $250-million industry—an industry that is expected to double in size in the next ten years.

Jan knows that the era of unrestricted collecting is, and should be, over. She welcomed the creation of quotas and the Carl Shuster Horseshoe Crab Reserve. She realizes that drastic measures must be taken to protect horseshoe crabs for the future benefit of all mankind. It is refreshing to hear such wisdom and candor from someone in the industry.

CHAPTER 19

Numbers, Numbers, Numbers

[Delaware Bay, September 28, 2001]

AS THE YEAR drew to a close, the future of horseshoe crabs seemed to be hanging in balance. United action by the fifteen Atlantic states had reduced the catch of horseshoe crabs from 3 million in 1998 to 1.6 million in 2000, an impressive 46 percent drop. The 1,500-square-mile Carl Shuster sanctuary off Delaware Bay and the smaller reserve on Pleasant Bay had been created and tested in the federal courts. Increasing use of bait bags had cut the number of crabs used for bait in some states by as much as one-half. Although the Associates of Cape Cod tossed out their hemolymph and announced they had stopped doing research on a lysate by-product or an artificial bait to replace horseshoe crabs, research continued at the University of Delaware. If it proved successful, bait made from lysate by-product could significantly reduce the pressure on live crabs. Above all, economics continued to make it nonsensical to use horseshoe crabs for bait. Why would anyone want to chop up a dead crab and sell it for $1 if he could sell the same crab alive for $10?

That's how things stood by the end of the summer of 2001. Environmentalists had hoped that the various measures had been taken in time. If so, it would mean that the number of horseshoe crabs would remain depressed for ten years, until new year-classes of crabs could mature and start spawning. Only then would the population start rebounding, during the following decade. That's the trouble with horseshoe crabs. They mature slowly, and all of the adult year-classes have been overfished. It looked as if it would take twenty years for the stocks to recover fully. As more numbers started coming in, however, the situation looked even worse than what everyone had expected.

Delaware Bay is the center of the region between North Carolina and New Jersey thought to contain 90 percent of all the horseshoe crabs that live on the East Coast. It is also the area where over half of the horseshoe crabs caught on the East Coast spawn. That is the reason scientists have spent the last decade going out during the high tides of spring to count the number of horseshoe crabs spawning on Delaware Bay. So far, they have chronicled a steady decline from 1.2 million crabs spawning in 1990 to less than 500,000 in 1999. More ominously, in 1999 researchers counted only 140,000 crabs spawning on the New Jersey side of the bay, formerly its most prolific area.

At the same time, ornithologists were reporting that shorebirds hadn't been finding enough horseshoe crab eggs on Delaware Bay to fuel their migrations north. They were arriving above the Arctic Circle too emaciated to start laying eggs large enough to hatch viable chicks.

Jim Cooper was dismayed. "I had hoped we had caught the situation in time. But now I am more pessimistic than ever before."

It is the same story that has been repeated in so many other fisheries so many times before. By the time scientists can collect enough data and managers can amass enough political will to act, the fishery has already crashed. But horseshoe crabs are different. In other fisheries, fishermen can always start catching another species, and consumers can switch to eating other kinds of fish. But no other animal can replace horseshoe crabs. They are the only source of *Limulus* amoebocyte lysate.

That fact presents another problem: no other procedure can replace catching and bleeding the crabs. But can it at least be done more efficiently? That would require better regulation. Unfortunately, although regulators have made several strides forward in regulating the bait industry, they have taken several steps backward in regulating the lysate industry.

Meanwhile the numbers of crabs used for lysate increased from 300,000 in 1999 to 378,000 in 2000. Some of this increase came about because of changes in the way horseshoe crabs were being caught. For example, the Associates of Cape Cod switched from relying on Jay Harrington, who caught horseshoe crabs by hand, to Rhode Island and Virginia fishermen, who caught their crabs by trawler. To their credit, most hand-collectors traditionally have treated their animals with care and resisted pressure to fish offshore or become middlemen between other fishermen and the lysate companies. However, the increasing demands of the industry have passed by the more careful methods of hand-collectors. After switching to trawlers, companies were reporting a 10 percent increase in the number of crabs they had to reject because of injuries sustained in trawl nets and trucks. Offshore fishing and long-distance transportation were taking their toll.

But collecting methods were only part of the story. Jim Cooper wrote that, based on the demands of the pharmaceutical industry, the lysate industry would need to double the number of crabs it bled in ten years. This will put the lysate industry on a collision course with the bait industry. By the time Dr. Cooper's article appeared, Massachusetts and South Carolina were already using more crabs for lysate than bait, and New Jersey, Maryland, and Virginia were using almost half as many. In ten years the total number of horseshoe crabs used for lysate would equal if not surpass the number of crabs used for bait.

Something else was also going on. Both Seikagaku in Tokyo and Haemechem in St. Louis were looking for land in Chincoteague. There was speculation on the island that both companies wanted to purchase their own raw lysate facilities. This could be a welcome development, because raw lysate operations reduce bleeding mortality and can stop the practice of transporting crabs

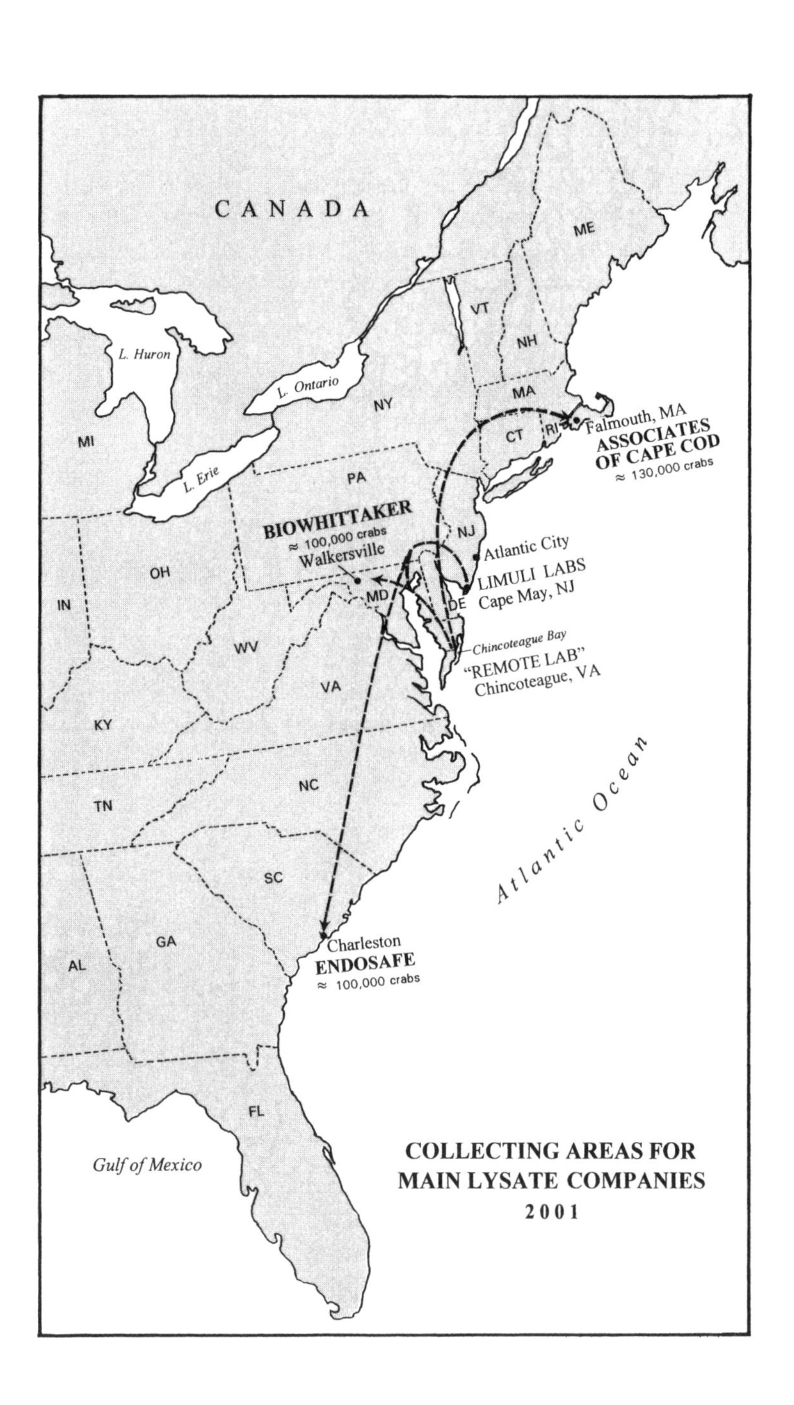
CANADA
ME
VT
NH
MA
NY
CT
RI
L. Huron
L. Ontario
L. Erie
MI
PA
NJ
OH
IN
MD
DE
WV
VA
KY
NC
TN
SC
GA
AL
FL
Falmouth, MA
ASSOCIATES
OF CAPE COD
≈ 130,000 crabs
BIOWHITTAKER
≈ 100,000 crabs
Walkersville
Atlantic City
LIMULI LABS
Cape May, NJ
Chincoteague Bay
"REMOTE LAB"
Chincoteague, VA
Atlantic Ocean
Charleston
ENDOSAFE
≈ 100,000 crabs
Gulf of Mexico
COLLECTING AREAS FOR
MAIN LYSATE COMPANIES
2001

to distant locations. But it would also mean that all the lysate companies eventually would be collecting horseshoe crabs between Cape May, New Jersey, and Chincoteague, Virginia—a strip of coast only a hundred miles long.

So what will the lysate industry look like in ten years? The Delaware Bay and Pleasant Bay reserves will protect major crab breeding stocks. They may be joined by sanctuaries off Monomoy and Chincoteague once sufficient research is done. And what will happen with those remaining stocks? About a tenth of the horseshoe crabs could be removed from the wild, bled, and returned offshore every year. It will be the same kind of ocean ranching that has gone on in Pleasant Bay for almost thirty years.

PART IV

Crabs and COVID

CHAPTER 20

Dr. Ling Jaek Ding

[Singapore, 2013]

THE LYSATE INDUSTRY continued to grow after I finished writing the first edition of *Crab Wars*. The five major producers were bleeding several hundred thousand crabs a year and making tens of millions of dollars. But the bait fishery was a different story. The Carl Shuster Horseshoe Crab Sanctuary had come too late. During the 1990s bait fishermen like Charles Burke and Bobby Bateman had removed over a million crabs a year out of Delaware Bay and birders had started to take notice. Red knots used to be able to gorge down 400,000 horseshoe crab eggs apiece during their two-week stopover on Delaware Bay. This would give them the fuel they needed to finish their migration from Tierra del Fuego to their breeding grounds above the Arctic Circle.

That had been easy when Delaware Bay's beaches were saturated with 100,000 eggs over every square yard back in 1991. But by 2000 there were less than 6,000 eggs per square yard and ornithologists observed that when the red knots arrived at their nesting areas, they were too emaciated to lay their eggs. This caused the population of red knots to plummet, so the U.S. Fish and Wildlife Service listed them as threatened under the Endangered Species Act. Environmental activists looked around and felt that maybe an artificially produced lysate might solve the bird problem. That idea had been around for almost as long as the lysate industry itself.

In the late 1970s horseshoe crab researchers heard rumors that somebody, somewhere, was going to make synthetic lysate and the LAL industry would go belly up. In the 1980s the rumors became more focused, saying that someone, somewhere in Asia was working on making synthetic lysate. In the 1990s we heard that somebody in Singapore was trying to use genetic engineering to produce an alternative to lysate. Finally, in 2003 we learned that the molecular biologist Dr. Ling Jaek Ding from the National University of Singapore had patented a formula for making synthetic lysate and licensed it to the Swiss firm Lonza.

It had taken 16 long years to get to that point. In the mid-1980s one of Ling Jaek's colleagues in the university's in-vitro fertilization lab had asked her to look into whether it was bacteria that were killing his embryos. Ling decided that the thousand-dollar LAL kit then available to her was far too expensive, so she decided to try to make her own lysate from the small round-tailed horseshoe crabs found in Singapore's mangrove swamps. This self-styled lab rat found herself and her husband Bow Ho up to their keisters in Singapore's pungent marshes looking for specimens of *Carcinoscopius rotundicauda*.

Ling Jaek soon learned that she couldn't extract enough blood from the small crabs without killing them. So she decided to try to use genetic engineering to make an alternative to LAL that wouldn't require using live crabs. She didn't have to start from scratch; scientists had already identified a molecule in the horseshoe crab amoebocyte blood cell that makes it able to detect bacterial endotoxins. She started to look for the gene that makes what scientists call Factor C. She finally discovered the gene, but she still had to figure out a way to splice it into another organism's cell so it would produce large quantities of recombinant Factor C, or rFC for short.

First she tried yeast cells, then mammalian cells, but neither of them worked cleanly. Finally she used a virus as a Trojan horse to insert the Factor C gene into the cell of an insect gut where, voila, it produced copious quantities of Factor C. This was because insects and horseshoe crabs share their lineage through arthropods, which is much closer than the lineage they share with yeasts and mammals.

After 16 years Dr. Ling finally had her alternative, so she sequestered herself in the Singapore University library and figured out how to license her formula and sold it to Lonza in 2003. Then she sat back and waited for something to happen.

Nothing happened.

Pharmaceutical companies are notoriously reluctant to change, particularly if it means they could become reliant on a single source of a new product that affects their manufacturing procedures. To say nothing of the fact that the companies making LAL could go broke if the Factor C alternative was adopted, and collectively they were pretty good at lobbying health and environmental officials on both the state and federal levels.

By 2013 things had started to change. When the pharmaceutical giant Eli Lilly decided to manufacture insulin in their plant in China, their toxin detection expert, Jay Bolden, convinced the company to use rFC to test all the products they made in China. Then they could simply present their data to the U.S. Pharmacopeia, a nonprofit organization that publishes a standard compendium of drug information, so the FDA could compare rFC with data from the standard LAL test for approval.

In 2018, Dr. Bolden planned to announce Eli Lilly's decision at a spring meeting of environmental organizations in Cape May, New Jersey. These would be his people. In addition to being an expert in toxin detection, Bolden was also an avid birder who knew that the world's population of red knots had been steadily declining because so many horseshoe crabs had been harvested for both bait and lysate. The feds had already banned harvesting horseshoe crabs for bait off Delaware Bay's Carl Shuster Horseshoe Crab Sanctuary, and now birders felt they could eliminate harvesting horseshoe crabs for lysate as well.

Everyone was excited. Birders felt like they finally had an ally in the industry and it looked like both the U.S. Pharmacopeia and the FDA would decide in favor of recombinant Factor C and the so-called blood harvest would come to an end. But everyone was in for a surprise—three big ones to be exact.

CHAPTER 21

Three Surprises

[May 29, 2020]

THE FIRST SURPRISE came in the guise of a powerful new ally. The social entrepreneur Ryan Phelan lived on a houseboat in Sausalito with her husband Stewart Brand, the 1960s visionary who had started the *Whole Earth Catalogue*. In 2012 the wealthy couple founded Revive and Restore, a nonprofit organization dedicated to using gene splicing to bring back endangered species.

It felt a bit like the entrepreneurs had discovered this neat new, New Age technology and they wanted to find a problem they could solve with it. The first problem they came up with was to bring back wooly mammoths and hold them in a Siberian tourist attraction called "Pleistocene Park."

If there was ever a problem didn't need solving, this was it. Bring back an Arctic-adapted animal to a world that was getting hotter than it had been in the last 50 million years? One can imagine one or two mangy mammoths shuffling through rapidly melting permafrost ponds sadly searching for nonexistent herd mates. It would probably make more sense to bring back heat-adapted Velociraptors and a Tyrannosaurus Rex or two. And what could possibly go wrong with that?

After also looking at bringing back passenger pigeons, black-footed ferrets, and Prezewolski's horses, in 2020 Revive and Restore bore in on using Dr. Ling Jaek Ding's rFC to save red knots and horseshoe crabs. They helped create the Horseshoe Crab Recovery Coalition, made up of several mainstream environmental organizations, and now well-funded and influential thanks to Revive and Restore's largesse. The coalition's stated purpose was "to save horseshoe crabs and the birds and fish that depend on their eggs to survive."

No mention was made of saving the lives of humans, half a million of whom had already died from COVID-19. One industry scientist resigned from the coalition panel because he felt Revive and Restore was pushing an agenda and questioned whether it was more important to save horseshoe crabs or produce the best possible test for endotoxins. This would be particularly germane in the midst of a pandemic, when the purity of testing and vaccine production need to be ensured. But it was much sexier to deploy a new razzle-dazzle technology like gene splicing than to do something as mundane as banning the use of horseshoe crabs for anything but critically necessary biomedical purposes.

Such a ban would be far more effective than rFC because bait fisheries used 45 percent more crabs than the lysate industry and accounted for 90 percent of the total mortality of the invaluable arachnids. In the 1990s I had broached the idea of banning horseshoe crabs for bait in several newspaper columns and on radio talk shows, expecting to get major pushback from bait fishermen. I was surprised that most fishermen didn't have a problem with the idea: "No, it's not a problem. We'll just use spider crabs or something else instead." By 2020 similar bans were working well in South Carolina, on Cape Cod's Pleasant Bay, and in the Carl Shuster Horseshoe Crab Sanctuary off Delaware Bay. But that didn't really matter. The idea of using gene-splicing technology garnered all the attention. *Audubon* magazine, the *Atlantic Monthly* and the *New York Times* carried glowing stories about the shiny new high-tech solution.

The second surprise came when people heard about a new coronavirus that had emerged from someplace called Wuhan, China. But most Americans only started to pay real attention when a cruise ship largely full of American tourists had to be quarantined for 12 days in Yokohama, Japan. Even then, Americans thought the new illness was only going to disrupt their lives for a few short months. It soon became apparent that this was something entirely new that would alter their lives for years, and that scientists would have to develop successful vaccines and reliable antibody tests for the rapidly spreading contagion.

Nose and spit virus tests were fine, but blood-based antibody tests were the gold standard. They involved using syringes, vials, stoppers, and glass slides to see whether patients had built up antibodies in their blood from past COVID-19 infections. The second way to fight the disease would be to develop safe and effective vaccines, and by early May 2020 there were already several dozen candidates. All of those syringes, vials, stoppers, and glass slides used for the antibody tests, as well as all the containers and production fluids used to make each batch of vaccine, would have to be tested to make sure they are not contaminated with bacterial endotoxins. And the way to do that was either to use LAL or rFC when it was approved and adopted by the vaccine makers.

Since the companies making LAL were already having trouble harvesting enough crabs to meet their immediate needs, it seemed that the U.S. Pharmacopeia would rapidly issue quality guidelines so the two tests could be compared and the FDA could approve recombinant Factor C as well as LAL for pyrogen testing.

The biggest surprise came on May 29, 2020, when the U.S. Pharmacopeia announced that they had 30 years of data on the efficacy of LAL but only two years of data on rFC, so they couldn't issue adequate guidelines to compare the two tests.

"That just boggled my mind," said Ryan Phelan.

A spokesman for the Swiss firm Lonza, which produced both LAL and rFC and had also teamed up with Moderna to produce what they hoped would be the first vaccine to be approved retorted, "It would only take the combined daily production of the three major LAL producers to test 5 billion doses of vaccine."

Phelan countered that each manufacturer would have to use 10 times that amount of LAL in order to test every step of the manufacturing process for every dose of vaccine that went out the door. This statement was a little misleading, because vaccines are not tested dose by dose, but batch by batch, and each batch could consist of hundreds, even thousands of doses.

Eli Lilly's Jay Bolden said the Pharmacopeia's announcement would delay the FDA's final decision by three or four years, which was about the amount of time that some researchers thought the pandemic would last. Given the delay, Bolden said that Eli Lilly would start using the European Pharmacopeia's guidelines, which he felt the FDA would accept because that would put rFC and LAL on an even playing field. Industry scientists felt that rFC would be less sensitive and more expensive to produce than LAL: less sensitive because rFC is only a single component of the crab's complex, multicomponent defense against bacterial infection; more expensive because it requires growing insect cells in rigorously sterile, precisely controlled thousand-gallon vats. But in 2018, Revive and Restore published results showing essentially identical sensitivity for the rFC and natural crab-blood assays for endotoxin, though industry scientists disputed this conclusion. Moreover, as of 2020, Lonza offers both rFC and LAL quantitative endotoxin-detection kits for nearly a identical price, about $2.50 per assay.

So, where do we stand today? The LAL test is one of the most commonly used procedures in modern medicine. It is used thousands of times a day, every day, in a range of settings—from major pharmaceutical companies to clinics in developing countries, and it has a long and distinguished history. When the U.S. government supplied vaccines against smallpox and anthrax to fight against bioterrorism in 2002, all the vaccines were tested with LAL made from horseshoe crab blood. At least 50,000 lives were saved in 2004 when LAL was used to determine that most of the world's supply of flu vaccine had been contaminated. And every year the lysate test protects the lives of a million humans, such as the people contaminated by E. coli in the wake of Hurricane Katrina. LAL has also been expanded to test for fungal diseases, and it was even used to ensure that the Mars rovers were free of earth-borne Gram-negative bacteria. And now horseshoe crab blood will be used to check that almost half a billion COVID antibody tests and vaccines are pyrogen free.

But the horseshoe crab blood test is the only commonly used medical procedure based on a single species of wild animal. And it is still chilling to think that that animal could be declining up and down the East Coast. So while it is likely that in a year or two when the immediate crisis of the COVID pandemic passes, the FDA will approve the use of rFC and pharmaceutical companies will gradually replace some of their usage of LAL with the rFC. This will reduce the 10 to 30 percent mortality of crabs used to make lysate, but it won't have any effect on the 100 percent mortality of crabs used as bait to fish for eels and conch. There is a much simpler and more mundane way to deal with that problem, and it draws from our experience with seafood.

For example, if we were to lose all the lobsters, striped bass, or shrimp on the East Coast, it would be an environmental tragedy. But if the East Coast were to lose all its horseshoe crabs, it would be a major medical disaster. While we protect our valuable food species with stringent regulations, to date we have only started to protect our far more valuable populations of horseshoe crabs. Thus, it is time that all the East Coast states follow the lead of South Carolina, and only allow horseshoe crabs to be used for biomedical purposes. They should also require that the lysate collectors prescreen crabs for injuries and transport them in moisture- and temperature-controlled trucks and return the crabs to the wild alive, safe, and unharmed. In the wild they should be separated on release to avoid rebleeding.

Finally, states should raise the cost of licenses to collect horseshoe crabs and impose higher fines on people caught with dead horseshoe crabs. These fees would help provide money for town and state officials to enforce these stricter regulations. Similar practices have made the lobster industry one of the best-managed and most lucrative of all fisheries. Certainly, we can do the same for this now medically invaluable species.

CHAPTER 22

The New Kid on the Block: Pease Industrial Park

[Portsmouth, New Hampshire, July 2020]

IN JULY 2020 the Swiss firm Lonza started production of a potentially new COVID-19 vaccine at its plant in Portsmouth, New Hampshire. Labeled mRNA-1273, the vaccine was being fast-tracked by the federal government under Operation Warp Speed, the 10-billion-dollar project that had given Lonza and its partner Moderna in Cambridge, Massachusetts, permission to start small-scale production while its vaccine was still undergoing Phase 3 clinical trials. Lonza wouldn't reveal how many doses they were producing, but if the trials proved successful, they could start manufacturing 100 million doses a year, and half a billion doses by 2021.

All of these doses would have to be tested with the LAL horseshoe crab blood test or its synthetic derivative rFC after it won FDA approval. Lonza had planned ahead so it would be uniquely positioned to produce new vaccines, because it was the only company that produced both its own LAL and its own rFC diagnostic kits at its production facility in Walkersville, Maryland.

Though Lonza was the new kid on the block, it had a distinguished history. It had been established in the small Swiss town of Gampel in 1897, and taken its name from the Lonza River that it used to produce electricity to make chemicals and fertilizers. The company had entered the biotechnology sector in 1974, opening facilities in India's Genome Valley and in Pearland, Texas, while also forming partnerships with large pharmaceutical companies like Bristol Meyers and Glaxo-Smith Kline to manufacture their drugs. Part of Lonza's long-term strategy was to form partnerships with small start-up biotech firms like Moderna for the large-scale manufacture of their new drugs.

They opened their first production plant in Hopkinton, Massachusetts, but the FDA cited them for so many violations that Lonza's management decided to start over and build a new facility in an industrial park at the former Pease Air Force Base in Portsmouth, New Hampshire. Their Hopkinton experience was a reminder, if anyone needed it, that producing vaccines can be a very risky business. In 2000 they went ahead and installed three 20,000-liter stirred-batch bioreactors in the Portsmouth facility so they would be able to produce large quantities of vaccines and other pharmaceuticals.

Lonza's most consequential long-term strategy had been their purchase of BioWhittaker's raw lysate "remote lab" and their LAL production facility in Walkersville, Maryland, that had been previously sold to Cambrex in Chicago. This gave them their own immediate supply of LAL to test for endotoxins. Then, in 2013 they supplemented this capacity by purchasing Dr. Ding's license to produce rFC. This gave Lonza insurance that they would have a supply of rFC if it won FDA approval.

Most researchers think that rFC will win approval and it will eventually prove to be cheaper and more reliable than LAL, because it will not be dependent on a dwindling species of wild animals. Indeed, as of 2020 Lonza offers both rFC and LAL quantitative endotoxin-detection kits for a nearly identical price, about $2.50 per assay. Of course, there remains the problem of scaling up to the levels of rFC production that would be needed to test the huge demand emerging from a vaccine for COVID-19, along with the vials, syringes, and needles required for its use.

By July 2020, Moderna's mRNA-1273 had the lead in Operation Warp Speed's push to fast-track a new vaccine. It was already undergoing Phase 3 trials in a hundred research sites where 30,000 volunteers were receiving two shots of the vaccine or two shots of placebos 28 days apart. The research sites had been carefully selected to include areas where the incidence was high enough so the volunteers had a good chance of coming in contact with people already infected with COVID-19. Hopefully the vaccine would give half the volunteers immunity if they came in contact with the disease during their daily lives. As it turned out, the White House would have been a particularly good site to have tested out the vaccine.

If the Phase 3 trials were successful, Lonza was already on track to produce its 200 million doses per year and 500 million doses by 2021. The only fly in the ointment was that Moderna, as a small start-up firm, had never previously developed a vaccine and put it through the regulatory process for FDA approval.

Since Operation Warp Speed didn't want to put all its eggs in one basket, it also had plans to test potential vaccines by Pfizer, Johnson and Johnson, Novavax, and Oxford University. These vaccines would also all have to be tested with the LAL horseshoe crab blood test. Once the immediate crisis of the pandemic passed, the FDA would probably gradually allow LAL to be replaced with rFC. That would be a positive development for horseshoe crabs, but as previously noted, it would have no impact on the 100 percent mortality of the many more crabs used as bait to fish for conch and eels.

CHAPTER 23

On Pins and Needles: Operation Warp Speed

[August 15, 2020]

WHEN COVID-19 ARRIVED on America's shores in early 2020, health officials looked around and realized the United States was abysmally ill-prepared to deal with such a pandemic. Department of Health and Human Services scientist Rick Bright wrote a memo warning that there were only 15 million syringes in the U.S. strategic stockpile but we would need at least 450 million to vaccinate the entire country. He eventually resigned and filed a whistleblower complaint noting rumors that up to 400 shipping containers of syringes had been exported out of the country and warning that needles were the weak link in the supply chain.

Bright stressed that it would take two to three years to satisfy the vaccine needs of the country and that you needed twice as many syringes as people because the vaccines would have to be delivered in two shots. He also made himself unpopular and was transferred from his job heading up BARDA, the Biomedical Advanced Research and Development Authority, for disputing President Donald Trump's claims about the efficacy of hydroxychloroquine. But he had convinced the White House trade advisor Peter Navarro, who told the president that the U.S. would need 850 million needles and recommended that BARDA identify alternate vaccine delivery methods and ramp up their production.

It turned out that Health and Human Services (HHS) already had an ace up its sleeve. In 2018 their Assistant Secretary for Health Admiral Brett Giroir, attended a World Health Organization conference on primary care in Astana, Kazakhstan. During lunch he sat near Jay Walker, an edgy young billionaire who had started the Internet travel-booking site Priceline. Now the ever-entrepreneurial Walker had a new dream. He explained that he intended to radically improve the health and future of countries in the developing world.

He paused, reached into his pocket, and in a move reminiscent of Steve Jobs pulled out the prototype for a syringe that would come from the manufacturer already filled with vaccine. He explained that, "All you have to do is squeeze this soft plastic blister and the device will push vaccine into a patient's arm and record the drug, dose, location, and time of administration on a computer chip for later transmission."

"Wow, this is amazing," said Admiral Giroir, "How do you plan to market it in the U.S.?"

Walker explained that his company ApiJect wasn't planning to operate in the United States. But he appreciated the admiral's enthusiasm, later telling a reporter for Medicalxpress.com, "He was the first person, if not the only person at the event, who understood the revolutionary nature of the platform." One thing led to another and in January HHS approved a $10 million contract to research and develop ApiJect's revolutionary new device.

When the coronavirus struck a few weeks later, Peter Navarro wrote his memo to the White House coronavirus task force recommending that they direct BARDA to identify alternate vaccine delivery methods and ramp up production. Now, instead of a leisurely five-year plan to save the world, ApiJect found itself in a feverish sprint to produce 100 million devices by year's end, under a $138 million contract to save the United States. ApiJect had only made a thousand prototypes before, but for that kind of money it was willing to make a few million more.

The contract was announced in the Rose Garden as part of Operation Warp Speed. But just in case things didn't work out quite according to plan, HHS placed smaller orders with Marathon Medical and Retractable Technologies for an additional 320 million less expensive needles, and with Becton Dickinson for an additional 50 million traditional needles. However, there was a major fly in the ointment. Retractable Technologies produced 80 percent of its needles and syringes in China and Becton Dickinson, America's largest syringe producer, imported most of their components from China. All of their syringes had been tested with both LAL made from American horseshoe crabs and a similar substance, TAL, made from two species of Asian horseshoe crabs, *Tachypleus tridentatus* and *Tachypleus gigas*.

Unlike the population of *Limulus polyphemus* that resides almost entirely along the coast of the United States and is starting to be well regulated, the two Asian species live in waters of several nations that have lax regulations, if they have them at all. Harvesting practices was another problem. In the U.S. biomedical collectors are generally required to return the crabs to the wild after bleeding, to minimize their mortality. But in Asia, the male and female crabs are harvested in pairs, and then bled to death for lysate before being sold to feed humans and to make chitin from the crabs' shells. Thus, the TAL industry converts what could be a sustainable harvest into one with a hundred percent mortality. The two *Tachypleus* species are also suffering from habitat loss and pollution, which is further leading to their rapid depletion. All these problems raise the question of whether the U.S. population of horseshoe crabs will be able to absorb the increased demand for LAL when the 450-million-year-old *Tachypleus* species become commercially extinct.

CHAPTER 24

Three Asian Crabs

[August 2020]

THE EXISTENCE OF the two *Tachypleus* crabs and the smaller round-tailed mangrove crabs, used by Dr. Ling Jaek Ding to make the first Asian lysate, pointed to the longevity of these ancient arthropods. They also pointed to a misnomer. Technically Dr. Ding's initial lysate should have been called CAL because it was made from the *Carcinoscorpius* mangrove crabs, but now all the Asian lysate is called TAL because it is made from the blood of the larger *Tachypleus* species. But how did these 450-million-year-old arachnids survive and end up in their present locations? Fossil evidence shows that their earliest common ancestor lived along the shores of what was then the globe-girdling Tethys Sea, so they had a worldwide coastal distribution.

Since that time these seemingly defenseless creatures have survived earthquakes, volcanoes, asteroid impacts, and half dozen severe Ice Ages. They saw the earth bloom with flowering plants and trees. They watched amphibians crawl out of swampy waters to walk on land. They saw dinosaurs come and go and they saw a spindly-legged primate rise up to dominate the world in a most alarming manner. And it is not just a rhetorical flourish to say that horseshoe crabs watched dinosaurs come and go. There is a fossil site in Glen Rose, Texas, where you can see the tracks of a dinosaur that was wading through the waters of a shallow lagoon beside a horseshoe crab that was crawling along the sandy bottom, presumably preparing to emerge and lay her eggs.

She probably died because the lagoon had gone anoxic, a common occurrence in such fetid, poorly oxygenated waters. After she died hundreds of invertebrates swarmed over her shell and picked it clean. Then the larvae of serpulid worms settled out of the plankton and started to construct their calcareous shells that made a perfect mold of the now famous, but unfortunate arachnid. *Sic transit Gloria mundi*.

Just as importantly, ancestors of these ancient arthropods witnessed plate tectonics create and destroy continents with the wild abandon of the Hindu god Shiva. And it was those irksome continents and supercontinents with names like Gondwana and Pangaea that isolated populations of horseshoe crabs until they became separate breeding species living in slightly different niches along the western sides of both the Atlantic and Pacific oceans. To make matters even dicier, horseshoe crabs live on the ever-changing interface between land and water, so they have to deal with variations in water temperatures and salinities caused by the hourly fluctuations of the tides and the daily changes of the seasons.

But when you think about it, coasts are also remarkably stable. Continents have come and gone. Beaches have grown and eroded, marshes have thrived and then been buried by rising seas, but shores have been around since the dawn of our planet. Now we have seven major continents and separate Atlantic and Pacific ocean basins and horseshoe crabs live on the western sides of these oceans where the continental shelves are broad and the Gulf Stream and the Japanese Kuroshio current carry warm water north, creating long shores with the temperate shallow waters preferred by all the existing species of horseshoe crabs.

Today, *Limulus polyphemus* ranges from Maine to the Yucatan Peninsula but the two *Tachypleus* species and *Carcinioscorpius rundicaudata* are scattered throughout Asia in slightly different but sometimes overlapping niches. And while the *Tachypleus* species get to hang out on nice sandy beaches their smaller round-tailed cousins are relegated to crawling through the area's muddy mangrove swamps. It is this scattered distribution that has made the Asian crabs vulnerable to the harvesting habits of our own spindly-legged primate species.

The round-tailed crabs live along the marshy coasts of India, Bangladesh, Malaysia, Thailand, Cambodia, and Vietnam, while *Tachypleus tridentatus* resides along the coasts of Japan, China, Taiwan, the Philippines, and the islands of Borneo and Java. It is questionable whether the occupants of these countries get along any better than the occupants of the states of the Atlantic East Coast or the horseshoe crabs and dinosaurs of that early Cretaceous Texas lagoon, for that matter.

This scattered distribution will make it difficult for humans to create marine sanctuaries where the harvesting of horseshoe crabs is prohibited or limited for only biomedical purposes. The International Union for the Conservation of Nature has declared horseshoe crabs an endangered species, but in light of the present COVID pandemic isn't it humans who appear to be more ill-equipped to survive such a viral storm? Is it us, or horseshoe crabs, that are the more endangered species? I think the smart money would place their bets on the species that has been around for the last 450 million years.

EPILOGUE

Afterthoughts

[September 9, 2020]

JUST BEFORE THE nineteenth anniversary of 9/11, I drove to my family's house on Cape Cod's Pleasant Bay. The bay was where the lysate industry first started harvesting its horseshoe crabs and it remains an important collecting area for Seikagaku's Associates of Cape Cod. It was also where I first became enamored with horseshoe crabs, and pillbugs, and dare I say it, ticks. And though I was often bitten by dog ticks, I never got sick, and never found a deer tick.

But today we can no longer get together in our family home because of COVID-19 and my granddaughter can no longer play on the lawn because of Lyme disease. And the only way we can save ourselves from this pandemic is to use horseshoe crab blood to check that our vaccines and antibody tests are pyrogen free.

When I first started writing this book I found myself asking why I was spending so much time writing about horseshoe crabs when thousands of human beings had just died in the World Trade Center, when anthrax was being sent through the mails, and when U.S. troops were being vaccinated against biological weapons they expected to encounter when they invaded Iraq. As you will remember, John Steinbeck asked the same question when bombs were being dropped on Europe and he was writing about collecting starfish in the tidepools of the Sea of Cortez. He decided that, "None of it is important or all of it is."

I now know that both of us had been asking the wrong questions. It made little difference that we were writing about starfish and horseshoe crabs when thousands of humans had just died. Each organism is part of the complex unity of life I had felt so intuitively as a child. The real question raised by COVID-19 is not so much how it originated and propagated throughout the world, but how the unity of life is so indispensable to our very existence and future. Here is a disease of humans caused by a virus that evolved in a complex mix of bats, ticks, snakes, and pangolins, plus a ferret or another as-yet-unidentified laboratory organism, which can only be thwarted with the aid of a spiderlike arachnid that first crawled out of the ocean 450 million years ago.

Life itself started from such a virus-like piece of RNA that existed in the twilight world between the living and the dead. It evolved through the archaebacteria into nucleated organisms like horseshoe crabs, ticks, and spiders and on to bats, fish, snakes, and humans. Each organism is important in its own right and each organism is part of the unity of life. We live in this vast, wonderful, interconnected biosphere. It is what makes our existence possible. It will survive our depredations but we splice, dice, and tweak it at our own peril. So yes, if "None of it is important or all of it is," I will still stake my claim on the latter.

Index

LIBRARY OF CONGRESS CATALOGING-IN-PUBLICATION DATA

Names: Sargent, William, 1946– author.

Title: Crab wars : a tale of horseshoe crabs, ecology, and human health / William Sargent.

Description: Second edition. | Waltham, Massachusetts : Brandeis University Press, [2021] | Includes index. | Summary: "William Sargent presents a thoroughly accessible insider's guide to the discovery of the lysate test, the exploitation of the horseshoe crab at the hands of multinational pharmaceutical conglomerates, local fishing interests, and the legal and governmental wrangling over the creatures' ultimate fate. Updated to include the 2020 pandemic"— Provided by publisher.

Identifiers: LCCN 2021020990 (print) | LCCN 2021020991 (ebook) | ISBN 9781684580767 (paperback) | ISBN 9781684580774 (ebook)

Subjects: LCSH: Limulus polyphemus. | Limulus polyphemus—Research. | Limulus test. | Pharmaceutical industry—Corrupt practices.

Classification: LCC QL447.7. S27 2021 (print) | LCC QL447.7 (ebook) | DDC 333.95/5—dc23

LC record available at https://lccn.loc.gov/2021020990

LC ebook record available at https://lccn.loc.gov/2021020991